SUPER NAVI GATORS

Exploring the Wonders of How Animals Find Their Way

DAVID BARRIE

Illustrations by Neil Gower

THE EXPERIMENT

NEW YORK

SUPERNAVIGATORS: *Exploring the Wonders of How Animals Find Their Way*
Copyright © 2019 by David Barrie
Illustrations copyright © 2019 by Neil Gower

Originally published in Great Britain as *Incredible Journeys* by Hodder &
Stoughton in 2019. First published in North America by The Experiment,
LLC, in 2019.

The Experiment, LLC
220 East 23rd Street, Suite 600
New York, NY 10010-4658
theexperimentpublishing.com

Many of the designations used by manufacturers and sellers to distinguish their
products are claimed as trademarks. Where those designations appear in this
book and The Experiment was aware of a trademark claim, the designations
have been capitalized.

The Experiment's books are available at special discounts when purchased in
bulk for premiums and sales promotions as well as for fund-raising or
educational use. For details, contact us at info@theexperimentpublishing.com.

Library of Congress Cataloging-in-Publication Data available upon request

ISBN 978-1-61519-537-4
Ebook ISBN 978-1-61519-538-1

Cover design by Beth Bugler | Text design by Sarah Schneider
Author photograph by Miranda Barrie

Manufactured in the United States of America

First printing April 2019
10 9 8 7 6 5 4 3 2 1

To Mary

The thing hath been from the Creation of the World, but hath not so been explained as that the interior Beauty of it should be understood.

Thomas Traherne (c.1636–74)

Contents

Preface

Right now, a crow is flying past my window. It looks purposeful, heading off on some mission known only to itself. A bumblebee is methodically visiting flowers in the garden. A butterfly flaps quickly over the wall, careers around wildly, settles for an instant, and then flies on. A cat walks along the path and slips into the undergrowth, while overhead an airliner packed with people makes its descent toward Heathrow.

Just look around you. Animals, large and small, human and non-human, are on the move—everywhere. They may be looking for food, or a mate; they may be migrating to avoid the winter's cold or the summer's heat; or they may just be heading home. Some make journeys that span the globe, while others only putter around their neighborhood. But whether you are an arctic tern flying from one end of the earth to the other, or a desert ant dashing back to its nest with a dead fly in its jaws, you must be able to find your way. It is quite simply a matter of life and death.

When a wasp flies off on a hunting expedition, how does it find its nest again? How does a dung beetle roll its ball of dung in a straight line? After circling an entire ocean, what strange sense guides a sea turtle back to the beach where she was born to lay her eggs? When a pigeon is released hundreds of miles away from its loft, in a place it has never gone near before, how can it find its way home? And what about the indigenous peoples who still, in some parts of the world, make long and difficult journeys by sea and land without so much as a map or compass, let alone GPS?[1]

The first question I want to address in this book is simply this: How do animals—including humans—find their way around? As you will see, the answers are fascinating in themselves, but they prompt further questions that touch on our changing relationship with the world around us. We humans are abandoning the basic navigational skills on which we have relied for so long. We can now fix our position effortlessly and precisely, anywhere on the surface of the planet, without even thinking—at the press of a button. Does that matter? We do not yet know for sure, but in the concluding chapters I shall explore the issues that are at stake. They are important.

Before we get started, a few words about everyday navigational challenges may help to prepare the ground. So, consider for a moment how you cope when you arrive by airplane in a strange city.

Your first navigational task is to find the way from the aircraft through immigration to the baggage hall. Even this kind of indoor navigation can present problems, especially if your vision is impaired, but we can usually overcome them by following signs. Once you are sitting in the taxi or bus, you can relax and let the driver make the decisions.

On arriving at your hotel, you have to find the check-in desk and locate your room—again signs are a big help. In the morning you may want to explore the neighborhood on foot. The beguiling voice of your GPS-enabled cell phone could give you exact directions, but that is not real navigation: You are being told what to do.

If you are independent-minded and prefer to find your own way, you will probably reach for a printed map. The first practical challenge is to place your hotel on the map—in other words, to determine your position. Next you need to find the sights you want to visit and work out how to reach them and how long it will take. That means measuring distances and estimating your likely speed, which raises the issue of measuring time. Though it may not at first seem obvious, navigation is as much concerned with time as space.

That will do for journey planning. Now you face another problem: when leaving your hotel do you turn right or left? You need to know which way you are facing before you can set off on your journey. There are various ways of solving this key problem. You could refer to the compass built into your phone, but you could also orient yourself by working out what street you are on. Looking at the shadows to see where the sun is might help, too. Then, once you start walking, you will need to keep track of your progress by checking landmarks and street names against the map.

As you make more and more excursions, you will start to grasp the layout of the city—how each part connects with its neighbors. This is a matter of remembering landmarks and establishing the geometrical relationships between them. As we all know, some people are much better at finding their way around than others, but if you are adept at this kind of navigation, you will develop the confidence to make longer and more complicated journeys without even looking at the map, and instead of merely going to and from your hotel, you may start to follow routes that connect different areas of the city with each other. By now you will have acquired a *mental map* of the city.

But you may employ a very different navigational technique. Instead of using a map, you could simply follow your nose until you find something that interests you, all the while keeping a close eye on which way you are going and how far you have gone, so that you can safely find your way back to your hotel.

This process has been likened to the method employed by the legendary Greek hero Theseus. As he entered the Minotaur's labyrinth, he unwound the ball of thread given to him by Ariadne, and it was this "clue" that enabled him to retrace his steps after killing the monster. A ball of thread is not a very practical navigational tool in a busy modern city, so in practice, navigation without a map depends on close observation and memory.

The distinction between navigating with and without the help of a map is a crucial one, and it extends to nonhuman animals. Maps (whether physical or mental) offer great advantages, not east the possibility of constructing shortcuts that can save valuable time and energy, or making detours to avoid hazards and obstacles. Some animals do seem to use maps of some kind (though obviously they are not printed on paper), but this is hard to prove, and discovering how such maps work is trickier still. These are some of the deepest questions faced by scientists exploring the navigational abilities of animals.

The structure of this book reflects the distinction between non-map and map-based navigation. In the first part, I focus on how animals can navigate without maps, and in the second, I discuss the possible use of maps—of various kinds, by different animals—and the evidence for the existence of map-like representations of the world in their brains. In the final part, I consider what the science of animal navigation means to us.

Each chapter is separated from the next by a short passage in italics that introduces some example of animal navigation—usually a puzzling one—that could not be comfortably accommodated in the main narrative. These will, I hope, help to entertain the reader, while also revealing how many mysteries remain to be solved.

Animal navigation is a big, complicated area of research, and in a short book like this I can only highlight some of its main themes. It is far from being an exhaustive account of the subject, and because it is aimed at the general reader, rather than a specialist audience, I have avoided the use of technical terms as far as possible.

What I have written reflects not only my personal interests but also, in part, the encounters with scientists that have shaped the course of my research. I have focused my attention primarily on describing *what* animals do and *how*, rather than discussing *why*. Trying to answer that last question would provide enough material for several more books.

Finally, I must say a word about animal welfare.

Strict ethical rules govern the work of the scientists who work in the field of animal navigation (as in other areas of research), and all those I have interviewed take their responsibility to avoid inflicting suffering very seriously. But some of them nevertheless conduct experiments in which animals are harmed, and any account of the subject that ignored the results of their work would not only be incomplete but wildly misleading.

I strongly believe that we owe respect to our fellow creatures and that we must therefore avoid casually putting our needs ahead of theirs. Exactly how we decide what experiments on animals are justified is a difficult question, but at the very least we should do everything we can to ensure that we inflict no pain. To be honest, I am not at all sure that we yet know enough about animals like crustaceans and insects to be confident of our judgments on this score.

Some readers may feel that harming animals in the pursuit of knowledge can never be justified—in any circumstances. An ethical case can certainly be made for banning all harmful experimentation on animals, though I suspect that very few of us would be willing to live with the consequences—especially where medical science is concerned.

There is plenty of room for debate about the ethics of scientific research on animals and I certainly do not pretend to have all the answers. But it would surely be wrong to hold scientists to a higher standard than the rest of us.

PART 1

NAVIGATING WITHOUT MAPS

Chapter 1

MR. STEADMAN
AND THE MONARCH

A remarkable schoolteacher entered my life when I was seven years old. He taught mathematics, but took little notice of the syllabus, or the age of his pupils. A lesson from Mr. Steadman that started with the theory of Pythagoras might well take a detour through topology, before disappearing down the rabbit-hole of non-Euclidian geometry. These were the things that fascinated him and no doubt he thought it was good to stretch our minds.

Mr. Steadman was not only a mathematician, he was also an expert entomologist, and he ran a moth trap at my school during the summer months. For me, the beginning of the school day was an enthralling prospect because I could join him in examining the night's take before lessons began.

My school was at the edge of the New Forest, one of the best places in Britain for insects, and often the trap would be filled with fifty or even a hundred moths, quietly resting inside the box to which they had been lured during the night by a brilliant light. Some moths and butterflies, I learned, were not natives, but came only as summer visitors. A common catch was the silver "Y" moth, which—we now know—travels up in vast numbers from the Mediterranean to breed in Northern Europe every summer. Why these insects made such long journeys and how they found their way was then a complete mystery.

I was soon obsessed with lepidoptera and, to my mother's dismay, my bedroom at home filled up with nets, collecting boxes, setting boards, and the tall cages in which I raised caterpillars. Sometimes at night, I lay awake listening to the munching of my endlessly feeding captives and the faint patter of their tiny droppings (or "frass") falling among the leaves of their food-plants. When they had eaten their fill, they turned into pupae (or chrysalids), their fat bodies dissolving into an alchemical soup, out of which the adult moths magically coalesced. To watch them break out of their hard, dry carapaces, slowly stretch their damp, crumpled wings, and eventually launch into flight, was to witness a miracle of nature, no less wondrous for its modest scale.

My long-suffering mother took me to the Natural History Museum in London, where a helpful young curator took us behind the scenes. Unlocking an unmarked door, he showed us into a vast room filled with mahogany cabinets containing millions of moths and butterflies from around the world. He pointed out one big, exotic butterfly, which, he said, turned up—very occasionally—in England. It came not from Europe or even Africa, but North America. Even if it was helped on its way across the North Atlantic by the prevailing westerly winds, or perhaps hitched a ride on a ship, this was an extraordinary feat.

The wings of this butterfly can be as much as four inches across and they look like modernist stained-glass windows. Delicate black veins fan out across a bright orange ground that glows as if the sun were shining through it. The dark lines join a thicker black margin flecked, like the animal's head, with snow-white polka dots. You might call this butterfly gaudy, but its loud color scheme warns predators thinking of taking a bite that they might be making a bad mistake. It may well be packed with poisons absorbed from its food-plant, the milkweed, while still a caterpillar. The butterfly, familiar to every North American, is the monarch.

I shared my excitement with Mr. Steadman, who quietly placed an order with an entomological supplier for a monarch pupa. When I opened the package, I recognized at once what it contained: Here was my very own *Danaus plexippus*.

The pupa was a heavenly work of the jeweler's art, perhaps only an inch or so long. Encased in shiny, jade-green armor, it lay nestled in its cotton-wool bed, like a miniature Chinese emperor awaiting his rebirth. I could dimly discern the shape of the wings and the segments of what might one day be the adult insect's body. A line of tiny, metallic golden dots gleamed in a half-circle around the fattest part of the pupa, which was spangled here and there with other touches of gold. It was a beautiful thing—more so to my eyes than the splendid adult—but also disturbing, somehow alien. How could the depths of outer space offer greater wonders, when our own world is filled with such glorious strangeness?

I never saw the butterfly emerge; it died before reaching maturity. But by now the monarch and its extraordinary life history had caught my imagination.

Many years later, I saw my first live monarch among the sand dunes of Amagansett, not far from Montauk at the eastern tip of Long Island. It was late August, and this butterfly was flapping, along with unseen millions of others, steadily to the south and west. Its flight was a care-free dance. A few lazy wing beats gave it lift, and then it glided for a few seconds, slowly losing height before again putting on the power. But where was it going, and how on earth was it finding its way?

It was my search for answers to these questions that set me off on the path that led eventually to the writing of this book. I knew there would be surprises along the way, but I had no idea how many and varied they would be.

The earliest navigators

When I began my researches, I was thinking only of animals I could see—like insects, birds, reptiles, rats, people—but the first life forms that emerged on our planet were very small indeed, and they were the pioneers of animal navigation.

Earth was born about 4.56 billion years ago, the chance product of the union of wandering asteroids drawn to each other by the force of gravity. It was not a very hospitable place in those days; molten rock covered its entire surface. The first continents emerged as this ocean of magma began to cool and harden about 4.5 billion years ago, but there were no oceans, nor even any air.

For hundreds of millions of years, the young planet was bombarded by yet more asteroids, but these explosive encounters were not entirely destructive. They delivered the chemical ingredients that gave rise to the very first living things, as well as water.[1] By 3.9 billion years ago, the earth had begun to quiet down and deep beneath the early oceans, simple forms of life began to emerge around hydrothermal vents— superheated jets of mineral-laden water, which then, as now, billowed out from the seafloor.[2] Among them were the very first bacteria.

Though we usually associate these single-celled organisms with disease, the vast majority of bacteria are harmless, and many of them make vital contributions to our physical and even mental health. In order to survive, they have ways of moving toward things they need (like food) and away from things that are a danger to them (like excessive heat, acidity, or alkalinity).[3] Some of them have specialized means of propulsion, including microscopic motors that drive rotating filaments called *flagella*. This simplest form of navigation is known as *taxis*—from the Greek word for "ordering" or "arranging."

Some bacteria engage in a particularly surprising form of taxis. The so-called magnetotactic bacteria contain tiny magnetic particles that, when joined end-to-end, act like microscopic compass needles. These "needles" force the bacteria to align themselves with the earth's magnetic field and thereby help them find their way down to the oxygen-poor layers of water and sediment where they flourish. The needles found in bacteria from the northern hemisphere have the opposite polarity to those in the southern hemisphere. A simple example of the power of natural selection.

Fossilized bacteria are extremely hard to identify, but the remains of magnetotactic bacteria have been found in rocks that are hundreds of millions, perhaps even billions of years old. Though they count as the earliest magnetic navigators in the history of our planet, living examples were first found only in 1975.[4] Oddly enough, their discovery coincided with the first demonstrations of magnetic navigation in much more complex organisms—like birds.

Our closest relatives among single-celled organisms are lumbered with a tongue-twisting name: choanoflagellates. Slightly more complex than bacteria, they live in water and they sometimes gather together in colonies. Like us, they depend on oxygen, and they can not only detect very small differences in its concentration, but also actively swim toward a richer source—again using their flagella.[5]

Even more impressive are the brainless assemblies of single cells known, rather unappealingly, as slime molds. These simple organisms can slowly but surely ooze their way toward a supply of glucose hidden at the bottom of a U-shaped trap. To do so, they employ a simple kind of memory that enables them to avoid revisiting places they have already explored.[6] They are also adept at solving a problem that human designers find challenging: the construction of an efficient rail network.

Researchers found that one particular slime mold, when presented with lots of oat flakes arranged in a pattern mimicking the layout of cities around Tokyo, set about building a network of "tunnels" to distribute the nutrients they extracted from the flakes. Amazingly enough, the network eventually came to match the actual rail system around Tokyo. The slime mold achieves this feat first by creating tunnels that go in all directions, and then gradually pruning them, so that eventually only those carrying the largest volume of nutrients (read "passengers") are left.[7]

Moving up the scale of complexity, the oceans—especially those that surround the Arctic and Antarctic—are filled with vast numbers of much larger, but still small, multicellular organisms known as plankton. Many of these plants and animals are invisible to the naked eye, but they are often so numerous as to make the sea look like a rich *miso* soup. Blooms of plankton can even turn the whole sea a rusty red.

Creatures like these have no need to know exactly where they are, which makes sense as they are largely at the mercy of the ocean currents, but they are far from being passive. In order to find food to eat, or to avoid getting eaten themselves, many of the animal plankton (that include fish fry, small crustaceans and mollusks) move up and down the water column, from the dark depths to the surface and then back again, every dusk and dawn. And the plant planktons that generally stay near the surface, to benefit from higher light levels, will plunge downward if necessary to avoid damage from excessive exposure to damaging ultraviolet light.

The timing of these events depends on the ability of plankton to detect changing levels of sunlight, though during the months-long Arctic night, animal plankton switch over to a rhythm based on moonlight.[8] In some cases, there may be more to these processes than a simple response to varying light levels. Certain plankton start to move before they can detect any change and even when removed to a dark aquarium, they continue to make their vertical migrations for several days. This puzzling behavior seems to depend on some kind of internal "clock" that governs their movements.[9] The entire oceanic food chain ultimately depends on plankton, and their colossal daily migrations play a crucial part in the life of the whole planet.

Even simple worms have to find their way around and one of them—a standard laboratory animal called *Caenorhabditis elegans*—seems to make use of the earth's magnetic field to steer when it is burrowing underground.[10] And newts, some of which can find their way back to their home ponds from distances of up to seven and a half miles, make use of a magnetic compass.[11]

Box jellyfish—small, transparent animals that are infamous in tropical Australia for the agonizing stings they can deliver—have no brain, but they do have eyes, and they don't simply go with the flow. They swim actively and with a real sense of purpose, hunting down their prey. Bizarrely enough, they have no fewer than twenty-four eyes, of four different kinds.

Even more surprisingly, some of them can navigate using landmarks above the surface of the water. One particular species, that frequents Caribbean mangrove swamps has a group of eyes that always point upward, regardless of the orientation of the animal's body. Heavy crystals of gypsum in the tissue around each of these specialized eyes maintain this orientation.

Dan-Eric Nilsson, a biologist at the University of Lund in Sweden (one of the leading centers of research in animal navigation), wanted to find out what these upward-looking eyes were doing. So he and his team put the jellyfish into clear, open-topped tanks, lowered them into the sea close to a mangrove swamp, and then monitored their behavior with a video camera. When the tank was within sight of the edge of the mangrove canopy, but a few meters from its edge, the jellyfish repeatedly bumped up against the side of the tank that was closest to the trees, as if they were trying to get closer to them. But when the tank was moved further away, where the trees were no longer visible from below the water surface, the jellyfish swam around randomly.

It seems that the jellyfish use their upward-looking eyes to pick out the silhouettes of the mangrove trees. This enables them to stay in the shallow water, where the tiny animal plankton on which they prey tend to congregate—though they can only do so if they don't stray too far from the canopy edge.[12]

These are only a few examples of the extraordinary navigational abilities displayed by organisms that may seem at first sight to be quite simple.

An old Walt Disney movie called The Incredible Journey *tells the story of two dogs—a Labrador and an ancient Bull Terrier—and a Siamese cat, which have been left by their owners with a friend. Not understanding that their stay in the strange house is only meant to be temporary, the miserable animals decide to find their own way home, but this involves crossing 250 miles of Canadian wilderness. After hair-raising encounters with a bear and a lynx, a narrow escape from drowning and a painful encounter with a porcupine, the three animals are eventually reunited with their family.*

Skeptics might well dismiss this story as literally incredible, but perhaps they should think again. In 2016, a sheepdog called Pero ran away from his new home in the English Lake District and found his way back to his original owners in Wales. He covered a distance of 240 miles in only twelve days and arrived in good condition, completely unexpectedly. Pero had a microchip, so there could be no possibility of mistaken identity.[13]

Nobody knows how Pero managed this feat. It is, I suppose, just conceivable that he found his way home by some extraordinary sequence of lucky choices, but that is very hard to believe. The navigational skills of dogs and cats have received surprisingly little serious scientific attention, though according to a recent study, dogs prefer to face either north or south when they relieve themselves. So perhaps they have some kind of internal compass that helps them at least work out which way they are headed. If so, they will join a rapidly lengthening list of organisms that are able to sense the earth's magnetic field.[14] But a compass alone would not have enabled Pero to find his way home.

It is possible that Pero managed somehow to keep track of where he was being taken when he went to his new home in the Lake District. Was he then able to reconstruct his route? Perhaps his acute sense of smell played some part in the process.

Chapter 2

JIM LOVELL'S
MAGIC CARPET

C harles Darwin (1809–82) wrote that "Man still bears in his bodily frame the indelible stamp of his lowly origin,"[1] but even he might have been surprised to learn that our eyes share the same ancient ancestry as those of box jellyfish, squid, spiders and insects.[2]

The unforgiving test-bed of natural selection has, over hundreds of millions of years, given rise to the eyes and brains that enable us (and other animals) effortlessly to pick out the things we really need to see—and to remember them. Not only do eyes help animals find food and mates, and avoid dangers, but, unlike the other senses, they can also provide exceptionally detailed information about distant objects, as well as those that are near at hand. For many animals, they are the single most important navigational tool, and we humans use them all the time to find our way around.

By comparison with many other animals, the typical city-dwelling human is not a very talented navigator, but with practice, most of us can manage pretty well with the help of landmarks. Our visual memories are actually very good, when we apply ourselves. We can, for example, recognize at least 10,000 images we have seen only briefly once before.[3]

Even powerful computers struggle to compete. Enabling them to perform quite simple tasks of visual recognition has proved extremely difficult. A computer looking for matches between two pictures of your house—one taken on a sunny morning and the other at night in the rain—will struggle. The changing position of a shadow, or the sudden appearance of a brilliant reflection from a window, will be enough to throw it into hopeless confusion. Raw processing power is not the answer, or at least not the whole answer. A supercomputer will have difficulty with visual recognition tasks, unless—like us—it "learns" how to focus on the features that are stable and relevant, while ignoring all the visual "noise." "Machine vision" is still prone to simple errors that we would never make, as accidents involving driverless cars have demonstrated all too clearly.

We all know what landmarks are typically like—think of the Eiffel Tower or the Hollywood sign in Los Angeles—but they take many different and sometimes surprising forms. They can be as large as Lake Michigan or the Great Pyramid, or as small as a single footprint. A route can be marked deliberately by leaving a trail of pebbles (as in the old fairytale), or by cutting "blazes" in the bark of trees with a hatchet. The ball of thread that Ariadne gave to Theseus might be seen as a single, extended landmark that marked his route to safety.

In addition to identifying a goal, or serving as waypoints along a route, visual landmarks can also provide valuable directional information. Take the Statue of Liberty that overlooks New York harbor, for example. Because her figure is not symmetrical, you can tell the direction from which you are looking at her by the shape of her silhouette.

Obviously the most important characteristics of a good landmark are that it should stand out clearly, and stay put for long enough to be useful, but oddly enough it need not be a solid object.

In the film *Apollo 13*, the astronaut Jim Lovell, as played by Tom Hanks, is about to depart on his ill-fated lunar mission. Trying to reassure his anxious wife, he recalls how he had once, as a young naval pilot in the 1950s, flown a sortie from an aircraft carrier over the Sea of Japan. It was night and he was fast running out of fuel, and if he failed to locate his mother ship soon, he would have to ditch in the "big black ocean." But the carrier was showing no lights, his radar had failed, and the ship's homing beacon was being jammed accidentally by a local radio station.

When Lovell tried to turn on the cockpit light to consult a chart, his electrical system shorted out and he lost all of his instruments. Now in complete darkness, he began to think about ditching—a risky procedure, even in daylight. It must have been a very scary moment. Then, as he looked down at the sea, he saw a long, glowing "green carpet" of bioluminescent plankton, which marked the turbulent wake of the very ship he was seeking: "It was just leading me home." And if Lovell's cockpit lights had not failed he would never have spotted it.

There are still a few indigenous peoples who have not abandoned their traditional navigational skills. While the ocean-going mariners of the Pacific Islands make heavy use of the sun and stars, the Inuit people of the far north rely mainly on landmarks to find their way—for the simple reason that they cannot count on having clear skies. In some areas, such as the coast of Greenland, there is no shortage of imposing natural features that can be seen from a great distance: mountains, cliffs, glaciers, and fjords. But in regions where the landscape is more uniform, the Inuit build their own landmarks called *inukshuks*. These resemble human figures and are usually placed on high ground with their arms pointing toward the nearest shelter.

According to Claudio Aporta, an authority on Inuit culture who has made long overland journeys in the Arctic, experienced Inuit wayfinders know thousands of miles of trails and can recognize countless landmarks along the way. Perhaps the visual memory of the Inuit peoples is unusually retentive, but they also make full use of a faculty available to all of us, the spoken word:

> Since Inuit did not use maps to travel or to represent geographic information, this enormous corpus of data has been shared and transmitted orally and through the experience of travel since time immemorial.

These oral descriptions rely on "precise terminology to describe land and ice features, wind directions, snow and ice conditions, and place-names."

The journeys the Inuit undertake can be extremely tough. Long waits in fog and "white-outs" are not unusual, but for the older generation, who learned to navigate before the advent of GPS, "the idea of being lost or unable to find one's way [was] without basis in experience, language, or understanding."[4] They are totally at one with their surroundings and make the fullest possible use of every navigational clue that is available to them.

The same can be said of the Aboriginal people of the land we now call Australia. They first arrived there by sea some 50,000 years ago, and like the Inuit, they have developed sophisticated navigational skills based primarily on the use of landmarks, and can follow long routes across the outback with the help of long and complex songs.

These enable them to recognize the natural features they encounter along the way by evoking mythic images from the "Dreamtime." As one expert (European) observer eloquently put it, Aboriginal methods of navigation are characterized by "belief in a spiritual power laying hold of material things and ennobling them under a timeless purpose in which men feel they have a place."[5]

Although Western city-dwellers cannot hope to grasp the intimate relationships that exist between Aboriginal and Inuit peoples and the landscapes they inhabit, our own distant ancestors may well have employed similar navigational techniques. It is sad to think that these cannot now be recovered, and it is therefore all the more important that the knowledge of those who still have such extraordinary skills should not be lost.

Some people speak languages that force them to consider which way they are headed all the time.

The Aboriginal Guugu Yimithirr of Queensland—from whom Captain Cook (1728–79) apparently learned the word "kangaroo"— never use words like "left" or "right." They use only the points of the compass:

If Guugu Yimithirr speakers want someone to move over in the car to make room, they will say naga-naga manaayi, which means "move a bit to the east" . . . [and] When older speakers of Guugu Yimithirr were shown a short silent film on a television screen and then asked to describe the movements of the protagonists, their responses depended on the orientation of the television when they were watching. If the television was facing north and a man on the screen appeared to be approaching, the older men would say that the man was "coming northwards" . . . If you are reading a book facing north and a Guugu Yimithirr speaker wants you to skip ahead, he will say, "go further east," because the pages are flipped from east to west.[6]

As Guy Deutscher says:

If you have to know your bearings to understand the simplest things people say . . . you will develop the habit of calculating and remembering the cardinal directions at every second of your life. And as this habit of mind will be inculcated almost from infancy, it will soon become second nature, effortless and unconscious.[7]

These linguistic peculiarities probably reflect the special navigational demands faced by the Guugu Yimithirr. For them a constant awareness of their orientation—an awareness embedded in the very structure of their language—may have been essential to their survival.

The six-legged secrets of a Provençal garden

I have had a soft spot for the French entomologist, Jean-Henri Fabre (1823–1915), ever since I first discovered his books. His major work, *Souvenirs Entomologiques* ("Entomological Memories"), the first part of which appeared in 1879, became that most unusual publishing phenomenon: a bestseller all about arthropods. Not only did he write some of the most lyrical and entertaining descriptions of insect life in any language, but he was also a pioneer of animal navigation studies.

Fabre was far from being a conventional scholar, but his exceptional powers of observation were coupled with the curiosity, patience and ingenuity that are the hallmarks of a true scientist. He spent much of his life struggling to support a large family on his teacher's salary, working in Corsica and in various parts of Provence. Though Fabre is often described as being self-taught, actually he had close links with the scholarly world and obtained an undergraduate degree as well as a doctorate. He eventually resorted to writing school textbooks to supplement his income—an activity that proved lucrative, allowing him to give up teaching and devote himself to his researches.[8]

Fabre was fascinated by the insects and spiders that must then have been far more abundant in the fields and hills of Provence than they are today, and he was especially intrigued by digger wasps. These parasitic animals lay their eggs in burrows and provide for the larvae that hatch from them by laying in stores of paralyzed prey, on which they can feast at leisure: a macabre living larder. He observed that while provisioning their nests, the wasps often traveled surprisingly long distances, and was amazed to discover that they could still find their way home, even when he took them several miles away.

Knowing from other observations that its two antennae played a key role in the wasp's search for prey, he wondered whether its navigational abilities also depended on these sensory organs. So Fabre simply lopped them off to see what difference that would make. He was surprised to find that this drastic procedure had no effect on the wasps' homing ability, though presumably it left the unfortunate creatures hungry.[9]

Baffled by the wasps, Fabre shifted the focus of his research to the fierce red ants that lived in his large garden—a species that raids the nests of black garden ants and steals their young.[10] These would be much more tractable subjects, as they could easily be observed during their forays outside the nest. With the help of his six-year-old granddaughter, Lucie, Fabre conducted a series of simple, but groundbreaking experiments.

First Lucie—with admirable devotion to duty—stood watch over the red ant nest, waiting patiently for a raiding party to emerge. She then followed the column and marked its path with small white pebbles, just like the small boy in the fairytale, as Fabre observed.[11] Once the red ants had found a black ant nest to pillage, Lucie ran back to tell her grandfather.

Fabre knew that red ants always retraced their outward journey exactly, when returning with their prey, and he thought they might be guided by some kind of odor trail. To test this idea, he tried by various means to remove or mask whatever scent they might be following. First, he attempted to disrupt it by vigorously sweeping the surface of the ground. But the determined ants, having been briefly delayed, found their way again, either by forging ahead over the swept areas, or by going around them.

Fabre suspected that some traces of a trail might have survived his broom, so next he trained a hose on the path, in the hope of washing away any remaining smell. But again, the ants eventually made their way past the obstacle. And it was the same story when he applied menthol to a section of the path, in an attempt to cover up the hypothetical scent.

Fabre now began to think that the red ants might be relying on visual cues, short-sighted though they plainly were, rather than scents to retrace their path. Perhaps they were memorizing landmarks of some kind. To test this idea, Fabre altered the appearance of the ants' homeward path, first by laying sheets of newspaper across it, and on a later occasion, a layer of yellow sand—quite different in color from the surrounding grey soil. These interruptions gave the ants a good deal more difficulty, though they still managed to regain their nest.

Fabre found that the ants could retrace their route to a source of prey even after the passage of two or three days, but when he moved the ants to parts of the garden they had never visited before, they were completely disoriented. On the other hand, they had no trouble homing successfully from areas they already knew.

On the basis of these observations, Fabre concluded that the ants were relying on sight rather than olfaction to retrace their steps. Though Fabre was astonished that an animal this small was clever enough to do such a thing, he was convinced that the ants, like human navigators, were using visual landmarks to find their way. His home-spun methods might not meet modern standards of scientific rigor, but he was definitely on the right track.

∽

Like Fabre, the great Dutch field biologist, Niko Tinbergen (1907–88), was fascinated by the way in which digger wasps returned unerringly to their burrows after going on their lengthy foraging expeditions. To Tinbergen's eyes at least, the small entrances seemed very inconspicuous. How were the wasps locating them? He thought that the wasps might well be memorizing landmarks, so he placed a ring of pine cones around the nest entrance. When he surreptitiously moved the cones, he was delighted to discover that the returning wasps looked for the nest entrance in the new location.

But were the wasps drawn to landmarks of any size or shape, or were there particular visual characteristics that attracted their attention more than others? To address this issue, Tinbergen tried placing markers of varying kinds around the burrow. Once the wasps had departed, he created two artificial entrances, each of which was surrounded by markers of just one sort.

It turned out that the wasps were more strongly drawn to dark, three-dimensional markers than pale, flat ones. Similar experiments with honeybees have shown that on leaving a nectar-rich flower, they take careful note of the surrounding landscape, with a special emphasis on three-dimensional landmarks. The bees can even make use of the geomet-rical relationships between these landmarks, especially their distance from the flower, to help them find their way back.[12]

Chapter 3

A TANGLED HORROR

The sweat bee is a native of tropical America and its rather unat-tractive name is explained by the fact that it likes to lap up human perspiration. While the more familiar honeybee flies by day, the sweat bee goes out only at dusk and dawn; it is a crepuscular creature. The females live in the rainforest and make their nests in small, hollowed-out sticks concealed in the undergrowth. When they head off on a foraging expedition, they have to pick their way through dense vegetation (though it is also possible that they fly over the top of the canopy—nobody yet knows for sure) and, judging by the pollen they collect, they can travel at least 300 meters.

It gets dark quickly in the tropics, and the darkness in a rainforest is very dark indeed, as the foliage blocks out much of whatever light is available. The navigational task of the sweat bee would be difficult enough in broad daylight, but once the sun has set, the scarcity of photons makes it "particularly challenging."[1] Quite an understatement.

I traveled to the University of Lund in southern Sweden to meet the man whose team made these extraordinary discoveries: Eric Warrant. An enthusiastic and energetic Australian, who knows as much about insect vision as anyone, he was plainly delighted to dis-cover that I shared his love of six-legged animals.

In the course of our conversations Warrant explained that you can test the sensitivity of an individual photoreceptor cell in the eye of an animal by recording its response to a point of light of varying intensity. When the light is extremely dim nothing happens, but as it is gradually turned up, the cell will start to "fire" minute electrical signals. Using this technique it has been shown that some animals can detect *single* photons of light.

It is worth pausing for a moment to reflect on the significance of that statement. A photon is one of nature's fundamental particles, though puzzlingly it also behaves in a wave-like fashion. We are talking about something so exceedingly small that it is said to be point-like: in other words, it occupies no space at all. Nor does it have any mass. A photon does however travel very fast (at the speed of light) and it delivers a tiny bit of energy (the amount varies with its wavelength).

That the eyes of any animal are capable of detecting such a minute packet of energy is astounding, but the sweat bee is in a class of its own. It manages to make its way home through the jungle on the meager visual diet of just five photons a second for each of its photoreceptors. Its nocturnal navigational skills give Warrant goosebumps:

> It's just absurd, absolutely absurd that they can fly through that tangled horror, find flowers and then effortlessly find their way back and land with such incredible precision.

The extraordinary sensitivity of the sweat bee's compound eyes cannot by itself explain how they navigate so successfully in nearly total darkness. Something more is needed. The answer lies in specialized cells in their brains that "add up" the signals coming from their eyes. These allow them to make the fullest possible use of the very limited flow of information arriving from the world around them. The slow flight speeds of sweat bees, compared to bees that are active in the day, also allows more time for the operation of this "summation" process. Warrant thinks that the sweat bee may well use the very dim patterns created by the contrast between the forest canopy and the night sky as landmarks to guide her back to her nest (as is known to be the case with some rainforest-dwelling ants), though this remains to be established.

As she leaves her nest, the sweat bee performs an "orientation flight" during which she studiously turns back and views the entrance and its surroundings. When Warrant and his colleagues moved a bee's nest after she had flown off, they found that she returned to the exact spot where the nest had been, presumably guided by the surrounding landmarks.

To test this idea they displayed a piece of white card over her nest entrance before the bee departed, and while she was away placed it on a neighboring, abandoned bee's nest. On her return the bee, deceived by the card, entered the wrong nest—from which she quickly departed. She could only find her way back to her proper home when the scientists returned the card to its original position.[2] Plainly then, the homing process is not based on smell.

People tend to look down on fish—and not just because we dwell in the air above them. To our superficial gaze they seem cold, slimy, and frankly rather dim. Why else would they be silly enough to take a hook or swim into a net? But in this, as in so many of our prejudices, we simply betray our ignorance. They are much harder to study in the wild than land animals, so our ignorance about fish is still profound, but one thing is certain: They do not swim around at random, and landmarks of various different kinds feature prominently in their navigational toolkits.

Fish have a variety of senses at their disposal, some of which are quite alien to us. Their lateral line organ—a series of pressure-sensitive pores along their sides—is exquisitely sensitive to the slightest movements in the water around them. It is this that gives shoals of fish their extraordinary ability to change direction in unison.

The blind Mexican cave fish makes use of the pressure waves generated by its own motion through the water to detect the presence and location of objects around it. As it swims through the darkness, its lateral line picks up the distinctive reflections they produce, and the fish can learn to follow routes based on these liquid "landmarks."[3]

Other fish, including for example the Indian climbing perch, make use of visual landmarks. This species lives either in ponds or fast-moving streams. Researchers took fish from these two very different habitats and taught them to find a reward by navigating through a succession of narrow doorways in their tanks. At first, the stream-dwellers did rather better than their still-water cousins, but when a small plant was placed beside each aperture, the results were reversed: now the pond-dwellers came out on top.

It seems that fish living in fast-moving water take little notice of such impermanent objects as plants, because they get swept away too quickly to be of any use as landmarks. The pond fish, however, can count on most things staying put, so they have learned to pay much closer attention to them.[4]

Several different kinds of fish, including eels and sharks, are sensitive to electric fields and make use of electric landmarks. The weakly electric fish, for example, has a specialized organ that enables it to detect changes in the electric field that extends through the water around it. It is a nocturnal animal that lives at the bottom of African lakes and, like the Indian climbing perch, it can learn to find an aperture in a barrier marked with a landmark using this technique. But there is one big difference: It does so in complete darkness.[5]

Even insects sometimes make use of electrical information to locate things.

When you peel the plastic wrapper off a package, it often sticks to your hand and refuses to let go. You may also get a mild shock when you touch a metal surface, especially after walking over an artificial-fiber carpet. These curious effects are caused by the build-up of a static electric charge and, oddly enough, they play an important part in the ecologically vital process of flower pollination by bees.

Bumblebees can detect the static electric fields that surround flowers, and can even discriminate between flowers on the basis of the different electrical patterns they produce. The bees pick up these faint signals with the help of sensory hairs that are deflected by the electric fields around the flowers. They use this electric information to tell the difference between flowers that provide lots of nectar and the less generous ones.[6]

Clark's nutcracker

Birds can fly over long distances, so the navigational challenges they face are especially demanding, but they do have wonderful eyesight—as well as a variety of other navigational tools. Just as we might sometimes use GPS and sometimes a map to find our way around, birds switch back and forth between them opportunistically.

Disentangling the roles of the different mechanisms that birds employ has proved extremely difficult, and many uncertainties remain. This is an example of a much wider problem that affects all branches of behavioral science. Interpreting the results of experiments on complex animals is seldom straightforward. Consider intelligence tests on humans. If a young child scores badly, does that necessarily mean they are not very bright? Perhaps they were anxious, distracted, or even bored—or maybe the test was poorly designed.

Despite these problems, it is quite clear that visual recognition is a key part of the navigational toolkit used by birds. And one bird in particular is a prodigy of landmark usage.

Clark's nutcracker is a member of the highly intelligent crow family, and it lives in the high mountains of the North American West. It was first described by William Clark, companion of Meriwether Lewis, the man who led the legendary overland exploratory expedition from St. Louis to the Pacific and back, in the early years of the nineteenth century, making maps along the way.

Clark's nutcracker can only survive the long cold winters in the mountains by stashing seeds during the summer months, like a squirrel. Being very far from stupid, it does not put them all in one place; that would be far too risky, because other animals (including other nutcrackers) will steal them if they get the chance and, of course, the bird itself would starve if it failed to find its own cache.

But the scale and complexity of the nutcracker's food concealment operation is stupendous. It hides only a few seeds at a time at sites scattered over 100 square miles of country. Some it may bury on windswept slopes, some in dense forests, and some on the bleak mountain-tops. A single bird may hide more than 30,000 seeds in as many as 6,000 separate caches. The birds need to be able to remember these locations, over a period of many months. Their powers of recall, though not perfect, are very impressive, and certainly more than adequate to allow them to survive in the tough environment they inhabit.

The caching behavior of the nutcracker exemplifies an important general principle that is particularly relevant to navigation: evolution favors the emergence of systems that are "good enough" rather than perfect. Nature "selects" those characteristics that will enable the organism to live long enough to reproduce. There is no point in acquiring a more complex mechanism, if a simpler one will satisfactorily meet this basic requirement, especially when the price of doing so is having a much larger brain. Brains are very greedy consumers of energy, which means a lot more food is necessary to keep them going. It does not pay to have a bigger brain than you really need.

You may wonder whether smell plays some part in the nutcracker's astonishing behavior, but that seems not to be the case. Instead, the bird takes note of small-scale landmarks positioned around each cache and can remember the geometrical relationships between them.[7] In the wild, these landmarks might be stones or bushes, though when tested in the laboratory, the birds are happy enough to make use of man-made objects. When the researchers surreptitiously move the landmarks, while preserving the patterns they form, the birds often search in the place indicated by the shifted array.

But it seems there is more to this bird's cache-finding system than that. Recent work[8] suggests that the birds place greater reliance on larger, more distant landmarks. These would be easier to spot at a distance and, thanks to their size, would also be less subject to the effects of wind and weather.

It is not yet clear exactly what signs the birds attend to in the wild, but they probably take note of prominent features in the environment surrounding each cache—such as trees or large boulders—perhaps recording a kind of panoramic "snapshot" of the locale. Finding a cache is then probably a two-stage process. First the bird identifies the neighborhood, by some kind of image-matching process involving large-scale landscape elements, then it homes in on smaller objects closer to the store that help to determine its precise position.

For thousands of years, people have exploited the extraordinary homing abilities of the pigeon to send messages quickly, and often over long distances. The military have used pigeons at least since Roman times, and hundreds of thousands were employed by the various combatants in the Second World War alone. Some of them were even awarded gallantry medals for faithfully delivering messages under fire.

There is a legend that the Rothschild bank made a killing in 1815 because they received news by pigeon post of the outcome of the battle of Waterloo ahead of the markets. It is a good story, though apparently without foundation. The Rothschilds did, however, develop a system of communication using pigeons and it was up and running by the 1840s,[9] some years before the first electronic telegraph systems were operational.

Pigeons were extensively used when Paris was besieged by the Prussian army in 1870–71. The birds were taken out of the city aboard balloons, which landed when they were safely beyond the reach of the encircling enemy. The pigeons were then fed and rested, before returning under their own steam with microphoto messages for the beleaguered inhabitants.

Because pigeons are so easy to rear, and are prepared to fly long distances at almost any time (unlike most birds), they have long been used to test different theories about how birds find their way. Electronic tracking devices have in recent years enabled researchers to study their homing behavior in great detail. Not surprisingly, pigeons find landmarks very helpful, though they can also follow learned "compass" courses.[10]

Young homing pigeons spend a lot of time exploring the surroundings of their home loft and, in doing so, get to know the layout of the local landscape, often over quite wide areas. The "survey" information they acquire in this way is of no use to them if they find themselves in a neighborhood they have never visited before, but as soon as they return to familiar territory, they lock on to prominent landscape features like roads, railways, and rivers to help them. The routes pigeons follow in the closing stages of their journeys become habitual and are often not the most direct.[11] But we should not be superior; they are behaving like the millions of human commuters who, as creatures of habit, often do exactly the same thing.

Pigeons seem to find it easier to learn a new route when the landscape over which they are flying offers a bit of variety, though not too much.[12] As the lead author of the study, Richard Mann, put it:

> Looking at how quickly they memorize different routes, we see that visual landmarks play a key role. Pigeons have a harder time remembering routes when the landscape is too bland like a field, or too busy like a forest or dense urban area. The sweet spot is somewhere in-between; relatively open areas with hedges, trees or buildings dotted about. Boundaries between rural and urban areas are also good.[13]

Contrary to popular belief, bats are not blind and many of them have exceedingly good eyesight. Some migratory species travel thousands of miles, and the ability to identify distant landmarks is obviously vital to them.

A few years ago, Israeli scientists took fruit bats fitted with GPS trackers from their home cave to a crater in the desert some fifty-two miles away. Some of the bats were released at the bottom of the crater, and others high up on the rim itself. Though the location of the crater was unfamiliar to them all, most of the bats still managed to find their way home.

Both groups of bats homed with equal success, but they behaved quite differently from each other at the outset of their journeys. The ones released on the crater floor, who could not at first see the surrounding landscape, were disoriented and spent some time going around in circles before they headed for home, while the group released on the rim headed directly for their cave. The bats seemed to be making use of large-scale landmarks such as distant mountains, and fixing their position by reference to them, like a hiker with a map and compass.[14]

∽

The tiny blackpoll warbler heads south in the autumn from northeastern North America and travels all the way to the Caribbean, and sometimes even as far as Colombia and Venezuela. Although sightings on board ships suggested that the migrating birds followed a route that took them right out over the Atlantic, for a long time it was unclear how much time they spent flying over the sea. But the mystery has now been solved. Using extremely small tracking devices, scientists have recently shown that they can fly nonstop from Long Island to Hispaniola or Puerto Rico—a distance of 1,721 miles across the open ocean.

Even when fattened up ready for migration, blackpoll warblers typically weigh only about 17 grams (.60 ounces)—about the same as fifty standard aspirin tablets. Though the ruby-throated hummingbird, which weighs only 3–4 grams (.11–.14 ounces), is believed to fly across the Gulf of Mexico on its own extraordinary migratory journey, that is a distance of only 530 miles. As the authors of the research say, the nonstop transoceanic flight of the blackpoll warbler is "one of the most extraordinary migratory feats on the planet."[15]

Chapter 4

OF DESERT WARFARE
AND ANTS

S everal days out of Halifax, Nova Scotia, and hundreds of miles from land, I was sitting at the helm of a yacht headed for England, when a small brown bird appeared out of nowhere and perched precariously on the guardrail beside me. It was so exhausted that it made no attempt to fly away when I approached it. Unlike the fulmars that were effortlessly skimming past the yacht, the poor creature was obviously not at home on the ocean, but it refused the offer of food or water, and eventually it fluttered off hopelessly. It may well have been a blackpoll warbler that had been blown off course, or perhaps it had made a disastrous navigational error and set off in completely the wrong direction.

The first challenge for any navigator, whether human or not, is to make sure they are heading in the right direction. This process is known as *orientation*. Visual landmarks usually provide the necessary clues, but if you are in an unfamiliar place, or out on the open sea where none are available, you are going to need some kind of compass.

The sun is not always visible, but it reliably rises in the east and sets in the west, and when it reaches its highest point in the sky (at noon), it is always either due north or south of you—except sometimes in the tropics when it can be vertically overhead.[1] So, in theory at least, the sun could help you work out which way you were headed.

But using the sun as a compass is not straightforward. As the earth turns on its axis, the sun traces an arc across the sky, and the points on the horizon at which it rises and sets as well as the *height* of the path it follows depend on the time of year and your latitude. In the tropics, for example, the sun rises almost vertically in the morning and then descends equally vertically after noon. In mid-latitudes, by contrast, it follows a longer and lower path across the sky.[2] In polar regions, the sun either stays above the horizon (the "midnight sun"), or below it, for months at a time.

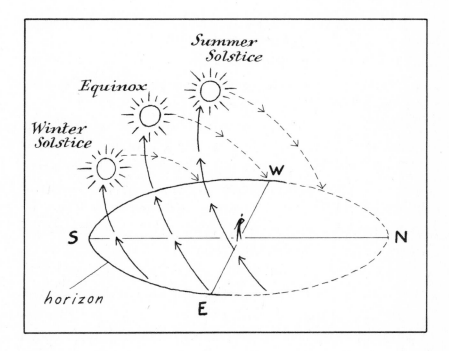

Typical paths of the sun during the day at a mid-latitude in the northern hemisphere.

The sun's progress across the sky is defined by its changing *azimuth*: This is the angle between true north and the point on the horizon that lies vertically beneath it.

Suppose you are in England in September and, like a swallow, you want to head south. What will happen if you set your course by the sun? Setting off at dawn with the sun on your left (azimuth 90 degrees), you will be heading in the right general direction. But as the day advances and the sun's azimuth gradually changes, your course will curve around to the right. By noon, when its azimuth is due south (180 degrees), you will be heading westward and by evening, as it sinks in the west, you will find yourself traveling in a northerly direction. You will in effect have followed a roughly U-shaped course—not a very satisfactory outcome.

Only by allowing for the sun's constantly changing azimuth can you hope to follow a steady course. But how is this to be done?

The answer is something called a time-compensated sun compass and, you may be surprised to learn, such a device influenced the course of the Second World War.

After the fall of France in 1940, the British army in Egypt was in grave danger of being overrun by a much larger Italian force based to the west, in Libya. The loss both of Egypt and the whole Middle East then seemed a real possibility. Without access to the Suez Canal and the oilfields of Iraq, Britain might well have faced defeat, and the Axis powers would then have been invincible. In that event, the world would now be a very different place.

Quite by chance, a remarkable man called Ralph Bagnold (1896–1990) happened to arrive in Cairo at this critical moment. A master navigator, he had explored the vast and then largely unmapped interior of the eastern Sahara Desert in the 1920s and 1930s using stripped-down Ford motor cars. Though only a major, Bagnold boldly ignored the "usual channels" and found a way of sending a memo direct to the new Commander-in-Chief, General Sir Archibald Wavell.

He recommended the establishment of patrols manned by specially trained volunteers that would penetrate deep behind enemy lines in "desert-worthy vehicles" to gather intelligence and conduct hit-and-run raids. Wavell immediately called for him and was much impressed by what Bagnold had to say.[3] With the general's full support, Bagnold very quickly found the men he needed and set up what became known, rather prosaically, as the Long Range Desert Group (LRDG).

Soon afterward, when the Italians began their eastward advance along the Mediterranean coast, the first LRDG patrols were secretly heading west through the desert, 300 miles to the south. The series of surprise attacks they launched had a dramatic effect: The Italians were so disconcerted that their advance stalled for several months. The delay gave the British time to reinforce and before long they were able to drive the Italian army back. The LRDG continued to play an important part in the later desert campaigns, but it was disbanded at the end of the war. Perhaps for that reason its remarkable achievements are less celebrated than those of the SAS, which was set up around the same time.

Accurate desert navigation was crucial to the success of the LRDG. The patrols depended on it for their very survival in the extremely testing conditions of the deep desert. But there was a problem: magnetic compasses were of little use to them. Not only did they react badly to the rough going, but they were also unreliable because the steel frames of the trucks gave rise to large errors. In fact, magnetic compasses could only be relied on when consulted some distance away from the vehicles. Since the patrols had to travel fast they could not afford to stop very often, so they badly needed something else to keep them heading in the right direction—something that worked well on board a bucking and rolling truck.

The answer came in the form of the simple, time-compensated sun compass that Bagnold had invented for his peacetime desert voyages. It consisted of an adjustable circular dial, marked around its edge in degrees, on which a vertical needle cast a shadow. A series of cards, one for each three degrees of latitude, showed the sun's azimuth at intervals throughout the day.

These were used to calibrate the compass, though around midday in summer the device was unusable, because the sun's shadow was too short to reach the scale around the edge of the dial. This provided a welcome excuse for the men to stop and shelter from the almost vertical sun. At night, the navigators could check their position by taking sights of the stars.[4]

Bagnold gave a vivid description of desert navigation using a sun compass in the account of his prewar expeditions:

> Our one thought was to keep awake and to keep the thin shadow on the dial of the sun compass on the arrow that marked the set course, for I knew that the little oasis would be difficult to find and so was anxious to hit it off exactly. For comparison, it was like starting from Newcastle on a compass bearing and trying to find a small garden somewhere in a vague rocky depression, which was the size of London and the same distance away [approximately 280 miles] . . .
>
> I had set the course so as to approach the oasis from the southwest . . . But it was all unfamiliar now; nothing fitted in with my memory of our previous approach. Our plotted position made [the oasis] eight to ten miles to the north-east, but on so long a run, we might easily be miles in error . . . In the half-darkness next morning only the vague outlines of nearby hillocks were visible. The wind came gently from the north-east, and I was distinctly aware of camels . . . [I therefore decided] to drive towards the smell, even though the country looked strange. After a few miles, I spotted the immediate surroundings of the oasis straight ahead.[5]

Since other animals do not have the navigational tables that Bagnold used to calibrate his sun compass, you might well think that they could not possibly steer by the sun. But you should never underestimate the power of natural selection, especially in the case of creatures that have been around for hundreds of millions of years.

The first hint that an animal might be using a sun compass emerged from the work of the British grandee and polymath Sir John Lubbock (1834–1913). Though a very different character from his near contemporary Fabre, Lubbock was also a pioneering investigator of the wonders of insect navigation. A banker, politician, archaeologist, anthropologist, and biologist, Lubbock was a close friend, neighbor, and devoted disciple of Charles Darwin. Though now almost forgotten, he was a well-known public figure in his day.

Lubbock was especially fond of ants and kept lots of them at his country house where, like Fabre, he explored their navigational abilities—though in a rather more formal way. Lucky weekend guests were treated to a tour of his beloved glass-sided ant colonies.

Lubbock wanted to find out how black garden ants found their way back to their nest. He first established that, unlike Fabre's red ants, they could follow a scent trail, but then he noticed something odd: the candles he was using to illuminate his work also seemed to be affecting their behavior. Puzzled, he conducted further experiments and eventually concluded that the orientation of the ants was "greatly influenced by the direction of the light."[6] Lubbock was too cautious to make a larger claim but, as later research revealed, the candles were plainly acting as proxies for the sun. This remarkable discovery was published in 1882.

A Swiss doctor in Tunisia

By the start of the twentieth century, a number of scientists were at work on ant navigation. Of these perhaps the most remarkable was an oddball Swiss physician from Lausanne called Felix Santschi (1872–1940), who came to Tunisia in 1901 at the age of twenty-nine and settled in the ancient walled city of Kairouan.[7] In this remote stronghold, the so-called Mecca of the Maghreb, he was to minister to the needs of the native people until shortly before his death.

As a young student in the 1890s, Santschi had accompanied a grand scientific expedition to South America, where he was gripped by a deep interest in ants. Living now on the edge of the Sahara, he was able to devote his spare time to observing and collecting the many different species that lived in the arid countryside. Before long Santschi began to publish scientific papers on the subject of ant navigation. His findings were groundbreaking, but because they appeared in obscure Swiss journals, they passed largely unnoticed at the time.

Santschi was an ingenious experimentalist who, unlike many of the leading scientists of his day, developed his theories on the basis of close observation of the *actual* behavior of animals in their natural habitat, rather than laboratory-based experiments built on assumptions about what they *ought* to do.

Despite Lubbock's discovery of the importance of light, the debate about ant navigation was still largely confined to discussion of the role that scent trails might play. Santschi, however, knew from his field observations that the desert ants he was interested in did not return to their nests along the same circuitous route they had followed on their outward journey. In fact, they followed a more or less straight course: an ant-line so to speak. In any case, the extreme heat meant that the volatile chemicals on which any scent trail must depend would evaporate too quickly to be of any use.

It was difficult to explain this remarkable behavior. A fellow student of the desert ant, a French civil engineer called Victor Cornetz (1864–1936), who was also based in North Africa, was baffled. He could only suggest that the ants were relying on an "absolute internal sense of direction," but he had no idea how such a mysterious mechanism might actually work. This did not satisfy Santschi, who asked a daring question.

Could the ants be using the sun as a compass?

Santschi came up with a simple but brilliant way of testing this novel idea. He set up a screen that would hide the sun from the ant, and then used a mirror to present its reflected image from the opposite direction. In most cases, the ant duly changed course by 180 degrees.

Whether or not Santschi was aware of Lubbock's earlier work, he deserves credit for being the first person to demonstrate that a sun compass played a part in an animal's navigational toolkit. Nor did he stop there. Santschi later showed that ants could navigate successfully in the twilight after the sun had set, and that they could do so during the day, when a cardboard cylinder (which he held over them as they walked along) allowed them to see only a small, empty circle of sky.

Santschi deduced that the ants did not need to see the actual disk of the sun in order to maintain a steady course. He found it difficult to explain these results, but he speculated that the ants might be making use of gradients in the intensity of light, or some other celestial cue—he even wondered whether they might somehow be able to see the patterns of the stars in daylight.[8]

Santschi's discoveries only received the recognition they deserved after his death, by which time similar behavior had been observed in honeybees.

The very first birds to be tracked using satellite technology were wandering albatrosses. These huge birds, which can weigh as much as 26 pounds, have been a source of wonder to sailors down the centuries, as they glide and soar effortlessly over the waves, scarcely ever needing to flap their giant wings. It was obvious, from the fact that they could follow ships for days or even weeks on end, that they could travel long distances.

But the true scale of the journeys they make only became clear in 1989, when two French scientists, Pierre Jouventin and Henri Weimerskirch, working in the remote Crozet Islands in the southern Indian Ocean, managed to attach satellite-tracking devices to six male birds during the breeding season.[9]

Laden with the 180-gram transmitters, the birds were returned to their nests, where they waited patiently until relieved by their mates. At this point, they headed out to sea searching for food. What the tracking devices revealed was breathtaking—and far exceeded earlier estimates.

One of the birds traveled over 9,300 miles in thirty-three days, another covered 6,479 miles in twenty-seven days, and one bird clocked 582 miles in a single day. They achieved average speeds as high as 36 miles per hour and in one case a maximum speed of 50 miles per hour. Riding the stormy winds of the Southern Ocean on wings that span three meters, these majestic birds have no trouble in circumnavigating the entire continent of Antarctica.

The albatrosses flew much further during the day than by night, stopping only occasionally, presumably to feed; but they kept going after dark, only much more slowly. It looks as if they are more comfortable navigating during the day, which may well mean that they rely at least partly on the sun for guidance.

Chapter 5

THE DANCING BEES

Together with Konrad Lorenz (1903–89) and Niko Tinbergen, Karl von Frisch (1886–1982) was one of the founding fathers of ethology: the scientific study of animal behavior in the wild. The extraordinary achievements of this indefatigable trio were recognized by the award of a shared Nobel Prize in 1973. Among these, perhaps the most impressive, and certainly the most famous, was the discovery of the honeybee's dance language, but it was a process that took many years.[1]

Honeybees explore the area around their hives in search of the nectar and pollen on which the hive depends for its livelihood, and their foraging activities may take them on journeys that cover as much as twelve miles. While studying how bees distinguish different flowers, von Frisch trained them to visit feeding dishes full of a sugar solution, which mimicked the nectar that fuels their long flights.

Then von Frisch made an intriguing observation. He noticed that the bees would from time to time return to a feeding dish that was empty, as if they were checking whether it had been replenished, and when he actually did top up the sugar solution, large numbers of bees started appearing at the feeding dish in a puzzlingly short space of time. It was as if they knew somehow what he had done.

In 1919 von Frisch borrowed a special hive that allowed him to watch what the bees were doing inside it—on the vertical surface of the honeycomb—through a glass panel. He trained a few bees to feed at a nearby dish and marked them with dots of red paint. He then allowed the sugar solution to run out before filling the dish up again.

Soon one of the trained bees, marked with paint, came to the dish and returned to the hive.

Observing the bee's behavior, von Frisch could not believe his eyes: It was "so delightful and riveting." Scurrying around on the surface of the comb, the bee waggled her abdomen, while the other bees excitedly turned their heads toward her and touched her abdomen with their antennae. If one of the marked bees was in the crowd, it would instantly set off for the feeding dish, but soon lots of the unmarked ones also started arriving there.

At first von Frisch suspected that the "recruits" were following a "scout" to the food source, but he was never able to find any evidence to support this theory. His thoughts, like those of Fabre and Lubbock before him, then turned to smell. So next he trained bees to feed from dishes placed on strongly scented surfaces—soaked with peppermint, for example, or bergamot—smells that would certainly be picked up on their feet and bodies.

The "recruited" bees showed a strong preference for food stations that were marked with these scents. Later von Frisch conducted similar experiments inside a greenhouse using real flowers, rather than dishes of sugar solution, with the same results. He concluded that the bee dances alerted the bees in the hive both to the *presence* of food and to its *quality*. He assumed, not unreasonably, that the recruits then located the new food supply simply by searching for the source of the scent they had detected on the dancer's body.

That bees could communicate with each other was a groundbreaking discovery. Although many scientists found it hard to believe that insects could be that sophisticated, the quality of von Frisch's work—and the brilliant lectures, books, and films with which he popularized it—had made him a world-famous figure by the time war broke out in 1939. But his reputation did not shield him from the unwelcome attentions of the Nazi regime. When someone revealed that von Frisch's great-grandparents had been Jewish converts to Christianity in the early nineteenth century, he fell afoul of the Nazi anti-Semitism decrees and came very close to losing his professorial post at the University of

Munich. He only managed to keep his job by undertaking to search for ways to increase honey production in support of the war effort.

Life was hard: By 1944, the Allied bombing raids had reached Munich and von Frisch saw his house and library destroyed, as well as his recently built laboratory. He was lucky to be able to take refuge with his family and some of his students at Brunnwinkl, his beautiful lakeside country estate in the foothills of the Austrian Alps, not far from Salzburg. The D-Day landings of June 1944, and the battles in northern France that followed, formed the grim background against which von Frisch and his colleagues began the momentous series of observations that forced him to alter, and vastly elaborate, his original theory about the significance of the bee dance.

The weather in August 1944 was ideal for bee work, and one of von Frisch's colleagues was conducting an experiment designed to encourage bees to make more honey, and pollinate more flowers, by leading them to better sources of nectar in more distant locations. Von Frisch suggested that she should condition the bees to go to a scented dish near their hive, and then remove the dish to the new, more distant site.

According to his long-held theory, the bees could be relied on to locate the dish in its new location simply by seeking out the source of the aroma they had learned to recognize. But he was in for a surprise: When the dish was moved, the bees failed to make an appearance and his colleague was left twiddling her thumbs.

During the course of that summer, von Frisch trained bees to feed at scented food sources, some of which were very close to the hive, while others were as much as 300 meters away. He found that when the scouts were trained to go to a distant food supply, their "recruits" would often fly directly to it—ignoring a much nearer one, even though it was marked with the same scent. This was very odd. Contrary to von Frisch's original theory, it appeared that the recruits were not merely searching for any food source that smelled right; they were actively seeking out a *distant* source and bypassing one that was closer to home. Von Frisch commented laconically in his notebook that the bees seemed to be capable of some kind of "distance communication."

When von Frisch ruled out the possibility that the bees were following an aerial scent trail, it became clear that the bees were indeed responding to distance information. They seemed, moreover, to show directional preferences, too. Could it be that the scouts' dances conveyed information, not only about the quality of a food source, but also its bearing and distance from the hive?

After the war, von Frisch worked hard trying to answer these fascinating questions. Using a paint-marking code that enabled him to identify large numbers of individual scouts, he showed that the speed with which they danced was indeed closely correlated with the distance of the food source they had just visited. And in the summer of 1945, he made some observations that were, if anything, even more surprising. The bees returning from one particular food source in the afternoon performed the straight segment of their waggle dance facing head down on the surface of the comb, but their orientation gradually altered over the course of the day—in line with the changing azimuth of the sun.

Von Frisch next explored how the orientation of the dance related to the positions of feeding stations, which he placed around the hive at the cardinal points of the compass: north, south, east, and west. What emerged was truly astounding. The dance direction consistently reflected the relationship between the bearing of the food source and the sun's azimuth. As von Frisch summarized his findings: "Dancing directly upward means: you must fly in the direction of the sun to get to the food source. Waggle dance pointing downward means exactly away from the sun is the path to the food."[2]

This was not only clear evidence of a form of celestial navigation in an insect, but also, more remarkable still, of the scout's ability to *communicate* information about the location of a food supply to its nest mates.

Von Frisch then placed a hive in a specially constructed hut, so that he could systematically manipulate the visual information available to the bees as they performed their waggle dances. He found that when sunlight was excluded from the hut (which was then illuminated for the observer's benefit with red light that the bees could not see), they were completely disoriented. But if he showed a flashlight, the bees would immediately direct their dances as if it was actually the sun—exactly like Lubbock's ants. And by moving the flashlight around, von Frisch could make the bees dance in any direction he chose.

Von Frisch then noticed that sometimes the bees could orient their dances correctly when they could only see a small sliver of sky. So, echoing Santschi's much earlier experiments with the desert ants (of which he was then unaware), he installed a stovepipe in the roof that restricted the bees' view of the sky to a narrow circle, in which the sun was not visible. So long as the sky was clear, the bees could dance correctly, but they became disoriented again when clouds crossed the circle of light. Next, he tried showing the bees a reflected image of the sky through the aperture of the stovepipe, and found that the orientation of their dances was then reversed.

When von Frisch discussed these puzzling findings with physicist colleagues, they came up with a possible explanation. They suggested that the bees might be sensitive to the polarization of sunlight.[3]

It had long been known that the light emitted by the sun consisted of electric and magnetic waves vibrating at right-angles to each other. Every possible orientation of these waves appears in sunlight as it travels through empty space, but when it passes through the earth's atmosphere, some of its components are filtered out. This process is known as polarization, and it results in the appearance of characteristic patterns in the sky known technically as "e-vectors" (short for "electric-vectors"). With our unaided eyes, we cannot see these patterns, but with the help of polarizing filters, we can get a rough idea of what they may look like to animals that can see them.

Try standing with your back to the sun on a cloudless morning, wearing polarizing sunglasses. If you look up at the sky overhead, you should be able to see a dark-blue bar running left to right, from horizon to horizon. If you then turn slowly 90 degrees, either to the left or right, you can see the dark bar gradually lighten. The dark bar marks the region of highest polarization, and its orientation in the sky is determined by the azimuth of the sun.

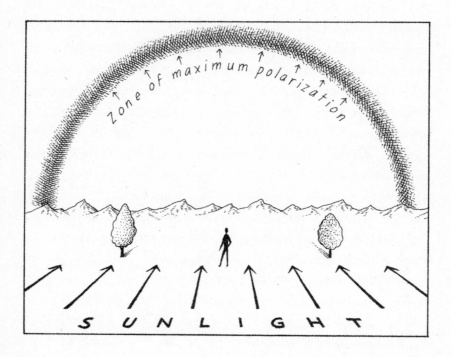

The band of strongest polarization in a cloudless sky when the sun is at your back.

Von Frisch realized that if the bees could see these patterns, they would not need to observe the sun itself; the e-vectors alone would enable them to determine its azimuth. And this he was soon able to show, with the help of polarizing films he obtained while on a lecture tour in the US from Edwin Land, the inventor of the Polaroid camera.[4]

Honeybee clocks

The discovery that honeybees could detect polarization patterns in the sky, and steered by them when the sun itself was invisible, was a major breakthrough, but simply knowing the sun's azimuth does not enable an animal to maintain a straight course—or at least not for very long. Somehow they must compensate for the sun's constant movement across the sky. And that means keeping track of time. Was it possible that, on top of all their other amazing talents, honeybees might also have a built-in clock?

A clue had emerged in 1929, but its significance was not immediately appreciated. One of von Frisch's students, Inge Beling, had then discovered that if honeybees were fed for several days at the same time, they would start turning up at the feeder on subsequent days at just the right moment to be fed. Later experiments showed that this remarkable behavior was not dependent on the availability of any external clues—such as the sun's changing azimuth. At the time, von Frisch wondered whether this mechanism was "a senseless gift of nature" or had some biological significance. It was not until the early 1950s that von Frisch was able to give a definitive answer to this question.[5]

Aided by his disciple, Martin Lindauer (1918–2008), von Frisch trained some bees to visit a food source some 180 meters away from their hive in a northwesterly direction during the afternoon, when the sun stood in the west. The following day, they transported the hive to a completely new location the bees had never visited before (so they could make no use of familiar landmarks for guidance).

They then placed dishes of food all around it at a distance of 180 meters—and in many different directions. Because it was morning, the sun stood in the eastern part of the sky. Nevertheless, the majority of the bees found their way to the dish that lay to the northwest of the hive, where they had been trained to look the day before. The only possible explanation was that they had made an allowance for the sun's changing azimuth.[6] Plainly this capacity depended on the bees' ability to keep track of the passage of time.

Confirmation that the bees had a time-compensated sun compass also came from another surprising source. When bees are about to swarm, they select the best location for their new nest by sending out scouts. When these scouts return to the hive, they perform dances that may last several hours, indicating the direction of the site they favor. Then other bees go out to inspect it and eventually, when a consensus is reached, the swarm heads off to its democratically selected new home. During the course of these marathon dance sessions, the orientation of the scouts' waggles changes in accordance with the changing azimuth of the sun. This would not be particularly impressive if they could see the sun or sky, but they adjust the direction of their dances even when they are inside a hive within a darkened room.[7]

Von Frisch's revelations about the navigational skills of the honeybee caused a sensation, because they seemed to imply that insects—despite their tiny size—were highly adaptable, and arguably perhaps even *intelligent*. For many of his scientific contemporaries, notions like these were extremely hard to swallow. They believed, as a matter of principle, that animals like bees simply could not be that sophisticated.

But there was also the problem that, in common with Tinbergen, von Frisch conducted most of his experiments outdoors, in a natural setting that could not be controlled as precisely as an indoor laboratory. The men in white coats seem to have found it difficult to take seriously the claims of a man who strode around the Alpine meadows wearing *lederhosen*. Perhaps their skepticism was tinged with envy.

Yet such was the rigor and elegance of von Frisch's work that most of the doubters were won over. A leading British ethologist of the day, William Thorpe (1902–86), who visited von Frisch soon after the war, commented in the scientific journal *Nature* that, "The zoologist may, indeed, be pardoned, if at first he feels skeptical—in spite of the immense detail and thoroughness of the investigation."

He mentioned one colleague who was almost "passionately unwilling" to accept von Frisch's findings, the implications of which were, he admitted, "certainly revolutionary." Thorpe himself was convinced, enthusiastically concluding that the behavior of the worker bee amounted to ". . . an elementary form of map-making and map-reading, a symbolic activity in which the direction and action of gravity[8] is a symbol of the direction and incidence of the sun's rays."

Although von Frisch's revised interpretation of the waggle dance steadily gained support, and indeed attracted interest far beyond the world of zoology, not everyone was convinced by his claims. Skepticism resurfaced toward the end of his career in a particularly alarming form when, in 1967, two young American researchers published the results of new experiments on honeybees—full of dense statistics—that directly challenged his key findings. Happily for the ageing scientist, fresh research that appeared in 1970 replicated his results and confirmed his conclusions.[9]

∽

With its slender, swept-back wings and gently dipping flight, the arctic tern enjoys a perpetual summer by commuting between the far north and the far south. But until very recently the scale of its seasonal journeys was not fully appreciated.

In June 2011, Dutch scientists caught seven arctic terns in the Netherlands and attached "geolocators" (weighing only 1.5 grams, or .05 ounces) to their legs. These devices recorded the time of sunrise and sunset every day—information that allowed the researchers to reconstruct the journeys the birds had made, when five of them were eventually recaptured a year later.

These birds had, on average, spent 273 days away from their colonies in the Netherlands and had traveled 56,000 miles. This counts (so far) as the longest avian migration ever recorded and it exceeds previous estimates for the same species by more than 12,000 miles. In an earlier study, terns from Greenland stayed mainly within the North and South Atlantic Oceans, following a roughly "figure-eight" course that took them down to the Antarctic and back. The birds from the Netherlands, by contrast, having reached the southern tip of Africa, then flew right across the Southern Ocean almost to Australia, before heading south to Antarctica, and then returning home via the Atlantic—a far longer circuit.[10]

Nobody yet knows for sure how the arctic tern navigates across the vast expanses of open ocean, or how it locates its breeding colonies.

Chapter 6

DEAD RECKONING

I t now seems astonishing that so many sailors were once willing to risk their lives crossing the oceans when the navigational tools at their disposal were so hopelessly inadequate. Imagine setting sail on a voyage that might last for several months without any reliable means of determining your position. Since fresh food could not be preserved and supplies of drinking water could be replenished only by rainfall, this was an even riskier undertaking than it would be today. Navigational shortcomings cost the lives of countless sailors, though they died more often of scurvy, thirst, or starvation than shipwreck. And as the exhausted blackpoll warbler revealed so clearly, we are not the only animals that have faced such problems.

In the distant past, open water navigation was such a hazardous business that most mariners probably clung to familiar routes whenever possible—though that certainly did not mean that they always hugged the coast. So long as they knew roughly how far they had to go, and in what direction, and could make a decent estimate of their speed and course, they could be fairly confident of reaching their goal. For navigators in the northern hemisphere, the height above the horizon of Polaris provided a handy measure of latitude, and from around 1500, thanks to the careful observations of astronomers, it also became possible to determine latitude by measuring the height of the sun at noon.

Provided the latitude of their destination was known, sailors could rely on reaching it—sooner or later—by sailing along it. But once they lost sight of land, they could not fix their exact position, because they had no means of determining their longitude. This meant that they could never be sure when they would arrive at their goal—a dangerous state of affairs, especially in heavy weather or when the visibility was poor.

The impossibility of measuring longitude also meant that there were no accurate charts. Estimates of the width of the Pacific Ocean, for example, varied by thousands of miles and the Solomon Islands—first stumbled on by the Spanish in the mid-sixteenth century—were then "lost" for two hundred years. Even in familiar European waters, charts were often wildly inaccurate. The "longitude problem" was not solved until the mid-eighteenth century, despite the offer of huge prizes by various different European governments over the previous two hundred years, and it was to be a long time before most mariners had access to the new technology and knew how to use it.[1]

How then did early mariners navigate on the high seas?

Leaving aside astronomical observations, they had three simple tools at their disposal. The magnetic compass (which seems to have entered use during the twelfth century in Europe), the chip log, and the lead line.

The compass, of course, provided a way of steering a steady course, though even this was not nearly as straightforward as it may sound, because these instruments were subject to a potentially dangerous effect called "deviation." This was caused by magnetic iron objects aboard the ship exerting an influence over the compass and, confusingly, it varied depending on which way the ship was headed.

It was not until the nineteenth century that this baffling problem was understood and effective remedies were developed. It also took a long time for people to grasp that sometimes there was a large difference between true and magnetic north, and that this varied not only from place to place, but also over time.

The chip log was simply a piece of wood at the end of a long rope, calibrated with knots tied at regular intervals. It was dropped over the side of the ship and allowed to run out astern for a fixed interval of time measured with a sandglass. The number of "knots" that ran out provided an estimate of the vessel's speed through the water. One knot was defined as one nautical mile (1.151 miles) an hour. This was a fairly effective system, though calibration of the log often gave rise to problems.

The lead line was, if anything, even less sophisticated. It was simply a long rope with a conical lump of lead on the end, which could be dropped over the side to measure the depth of water. By stuffing some fat into a cavity in the bottom of the lead, it was also possible to sample the composition of the sea floor (the "ground")—to see whether it consisted of sand, gravel, or mud, for example. Charts of coastal waters showed the nature of the ground and this information, combined with the depth, could be helpful in determining a ship's general position.

But, of course, the standard "lead" was of no use in the open ocean, where the depth of water typically reaches several thousand meters. Out there, old-time navigators could only estimate their position by the simple expedient of recording how long they had been going in a given direction. So, ten hours sailing at five knots on a westerly course meant that you were fifty miles farther west than you were ten hours earlier. Or so you hoped.

By recording each change of speed and direction (usually on a simple peg board, since most sailors were illiterate), it was in theory possible to work out where you were relative to your point of departure—even after a series of course and speed changes. This process is called "dead reckoning" (DR).[2] It is often said that DR means "deduced reckoning," but the term dates back at least as far as the seventeenth century, and its origins are quite mysterious. I like to think it was coined by an Elizabethan sailor with a dark sense of humor.

The trouble with DR is that it is unreliable; in fact, *very* unreliable. It is subject to many errors that are very hard to control. First, there are currents to contend with, and these may be strong, even in the deep ocean. There is no means of detecting them, unless you have some way of fixing your position. The log may say you are going at five knots, while the compass assures you that you are going west; but if the whole ocean is on the move, you may actually be heading in a different direction, and at a different speed. There is also the problem that sailing ships have a tendency to "sag" when the wind is not blowing from dead astern (directly behind the ship). In other words, they drift sideways, as well as going forward. Although it is possible to estimate the amount of this "leeway" by comparing the angle of the ship's wake with the course that is being steered, it is far from being a precise science.

Then there is the helmsman to consider. Some are good at keeping a ship on course, while others are less reliable. At the end of each watch, the navigator may be assured that the ship has been going steadily westward at a certain rate, but actually it may have followed a much more erratic path and its speed may also have varied. And, of course, there is always the weather. When a ship is driven before a storm it is impossible to keep track of anything, while in a flat calm it will simply drift at the mercy of invisible currents. In circumstances such as these DR breaks down completely.

Commodore Anson of the Royal Navy headed a famous expedition in the 1740s, which provided vivid illustrations of the unreliability of DR. Having struggled to round Cape Horn in appalling conditions, he believed that his small and battered fleet had made sufficient progress out into the Pacific that they could safely head north, up the west coast of South America. But he was in for a nasty surprise.

In the middle of the night, when Anson was sure they were far out to sea and well clear of the land, the lead ship fired a warning gun; they were headed for destruction on the rocky cliffs of Tierra del Fuego. It was a very narrow escape. Their DR was out by about 500 nautical miles (575 miles). Later, his first attempt to locate the Juan Fernández Islands failed and the delay cost the lives of dozens of sailors who died of scurvy.

Mark Twain goes in circles

In the 1950s, an entirely new navigational challenge emerged with the development of nuclear submarines that could operate submerged for months at a time. Although by then celestial navigation had long since been perfected, and various radio-based forms of position-finding were also available, these tools were not available to vessels patrolling deep beneath the surface of the sea.[3]

The answer came in the shape of a navigational system that recorded accelerations in three dimensions—in other words, changes in the speed and orientation of the vessel—with the help of an array of gyroscopes. By integrating the input from these inertial sensors a computer can track every maneuver the submarine makes and generate an accurate position at any given moment. Allowances, however, have to be made for the rotation of the earth itself, and the system needs to be updated from time to time, as it will otherwise gradually "drift." Inertial navigation, as this system is called, has been widely used in missiles, airliners, and even spacecraft.

Curiously enough, we humans employ a similar mechanism, as do many other vertebrates. It is known as the vestibular system. Our inner ears are designed to detect accelerations, like the gyroscopes on board a submarine, though they work in a different way. Tiny stones called otoliths, inside the semicircular canals, exert pressure on sensitive hairs, and these send signals to the brain, which can then work out the direction and speed of your body's movements. But that is not all. At the same time, you are receiving valuable feedback from your joints and muscles. For example, by counting the number of strides you have taken you can estimate the distance you have covered, and by feeling the slope of the ground and the effort required, you can judge whether you are going uphill or down.

By integrating information from these various "self-motion"[4] cues, in principle it ought to be possible for us to keep track of where we are. But sadly this system does not work very well in practice, as the following tale illustrates.

The world looks very different after a snowstorm. Many of the landmarks on which the traveler would usually rely are hidden, and without good local knowledge—or the skills of an Inuit hunter—they may soon get into trouble.

This is exactly what happened to the famous American author Mark Twain (1835–1910) when he and his companions were on their way to the frontier town of Carson City, Nevada, in the mid-nineteenth century.

In his autobiographical book, *Roughing It*, Twain describes how he and his party, which included a know-it-all Prussian called Ollendorff and a character called Ballou, almost came to a chilly end. Thick snow concealed the road, and because the visibility was poor the travelers could not set their course by the distant line of mountains:

> The case looked dubious, but Ollendorff said his instinct was as sensitive as any compass, and that he could "strike a bee-line" for Carson City and never diverge from it. He said that if he were to straggle a single point out of the true line, his instinct would assail him like an outraged conscience. Consequently, we dropped into his wake happy and content. For half an hour we poked along warily enough, but at the end of that time we came upon a fresh trail, and Ollendorff shouted proudly: "I knew I was as dead certain as a compass, boys! Here we are, right in somebody's tracks that will hunt the way for us without any trouble. Let's hurry up and join company with the party."

Twain and his companions put their horses into a trot and, finding that the tracks left by their predecessors were growing more distinct, deduced that they must be catching up. At the end of an hour, the tracks looked "still newer and fresher" and, rather surprisingly, the number of travelers ahead of them seemed to be steadily growing:

> We wondered how so large a party came to be traveling at such a time and in such a solitude. Somebody suggested that it must be a company of soldiers from the fort, and so we accepted that solution and jogged along a little faster still, for they could not be far off now. But the tracks still multiplied, and we began to think the platoon of soldiers was miraculously expanding into a regiment—Ballou said they had already increased to five

hundred! Presently he stopped his horse and said: "Boys, these are our own tracks, and we've actually been circusing round and round in a circle for more than two hours, out here in this blind desert! By George, this is perfectly hydraulic!"[5]

Literature and folklore are full of such stories, and they are borne out by scientific research, though there has been a good deal of argument about the causes.

Back in the 1920s, a scientist called A. A. Shaeffer claimed that human beings had a strange, innate "spiraling tendency" that kicked in automatically when we could no longer see where we were going. It was this, he argued, that caused us to "go in circles."[6] Others, however, claimed to have evidence that disparities in leg length, changes in posture, distractions, or errors in the placement of our feet (to name a few examples) might contribute to the failure of our internal navigation system.

Much more recently, Jan Souman conducted an experiment,[7] in which he asked his subjects to walk blindfolded across a large, flat airfield. There were no sounds to guide them, and he found that they were unable to keep a straight course—even over short distances. They followed tortuous and apparently random paths, and often circled back on themselves, with the result that on average the farthest distance any of them ever reached from their starting point was about 100 meters.

As far as Souman could tell, there was no pattern in these errors, nor was there any sign that physical influences such as unequal leg length or strength were to blame. Another researcher had earlier investigated how long people could maintain a steady course toward a target, after it was suddenly concealed from them. They could do so for only about eight seconds.[8]

Even when some visual information is available, our ability to maintain a straight course is quite poor—unless the sun or moon is shining. Souman tested people walking without blindfolds in two radically different environments, neither of which offered them many useful landmarks: a German forest and the Tunisian desert. The results were interestingly mixed.

In cloudy conditions, the subjects all had great difficulty in going straight, but when the sun was out they did much better, often maintaining a steady heading over surprisingly long distances, even in the cluttered and confusing environment of the forest. One subject walking by night in the Tunisian desert also did quite well, as long as he could see the moon. But when it was hidden behind clouds, he made several sharp turns and eventually headed back the way he had come.

These findings suggest that most people can steer by the light of the sun or moon by performing a rough-and-ready kind of time-compensation. But there are good reasons for our inability to maintain a constant course on the basis only of internal, self-motion signals. Systematic errors inevitably creep in, and they tend to accumulate. A directional bias is therefore bound to appear eventually. It follows that if an animal (of any kind) wants to stay on the straight and narrow, it must have access to external checks, whether in the form of landmarks or some form of compass. If not, its path will sooner or later approximate to a spiral.[9]

So perhaps Schaeffer was right all along—maybe we do have an innate spiraling tendency.

∽

In 2009, a land bird called the bar-tailed godwit was tracked flying nonstop across the Pacific Ocean all the way from Alaska to New Zealand in just over eight days—a distance of roughly 7,250 miles.[10] Several other birds followed routes that were only slightly shorter, so this was clearly not a freak occurrence. For a bird that has to flap its wings to generate lift—as opposed to soaring and gliding like a wandering albatross—to travel that far is almost incredible, and all the more impressive when you consider that godwits cannot land on water, because once wet, they can't get into the air again.

These extraordinarily long flights place massive physical demands on the godwits, which are forced to increase their resting metabolic rate by a factor of 8 to 10 solely to remain airborne. They then have to maintain *that level of exertion for the duration of the journey. In order to supply their energy needs, the birds fatten themselves up enormously before depar-ture, and their vital organs shrink to keep their take-off weight to a minimum. By the time they reach New Zealand—more dead than alive—they will have lost a third of their body weight.[11] But the birds also have to find their way across thousands of miles of empty ocean and cope with the effects of unfavorable weather en route. How they do so remains unclear, though it is interesting that they time their departure from Alaska carefully, in order to take advantage of tailwinds.[12]*

But why should the godwits choose to fly directly across the open ocean, when they could follow the continental margin of Asia? Several factors seem to be at work. It looks as if the direct route not only saves the birds valuable time, but minimizes the overall energy costs. Flying over the sea also enables them to avoid predators such as peregrine falcons, and reduces the risk of exposure to parasites and disease. However, the balance of advantage must be different when they head north again; this time they follow the coast for much of the way.

Any alterations in the seasonal winds over the Pacific, caused by climate change, will disrupt the bar-tailed godwit's transoceanic migration, and they are also threatened by the rapid loss of the wetlands in China, where they stop to refuel on their northward journey.

Chapter 7

THE RACEHORSE OF
THE INSECT WORLD

F or all its shortcomings, DR is the only practical way of keeping
track of your position—unless you have some independent means
of fixing your exact location, such as access to landmarks or GPS. And
over very short distances, before the various errors have had time to
accumulate, it can be quite effective. So it makes sense to ask whether
other animals can make use of DR. The fact that desert ants can follow
a complicated, zigzagging course on their foraging expeditions, and
then head straight for home, suggests that they may be promising
candidates. In order to find out more about the navigational abilities
of ants, I traveled to Zürich to meet the world's greatest expert on that
subject: Rüdiger Wehner.

Wehner's single-minded determination to understand the homing
behavior of the desert ant has been truly formidable. Like von Frisch,
he has conducted hundreds of experiments in the field, but he has
also employed the tools of neuroscience, anatomy, molecular biology,
and even robotics to explore the many different navigational mechanisms that enable the desert ant to flourish in a very unforgiving
environment. There is much talk in the world of science about the
value of interdisciplinary research, but few researchers have pursued
that ideal with as much determination and success as Wehner.

Although my train was due to arrive late at night, Wehner insisted
on meeting me at the central station in Zürich. His tall, bespectacled

figure was an unmistakable landmark in the middle of the vast, almost empty concourse. The following morning, after breakfast in the university canteen, we went to his apartment, which looks out over the lake to the high mountains in the west, and spent the whole day in his study discussing his work. Though most of the books that lined the walls were scientific ones, there were also many plays and novels, as well as works of philosophy and art history. Our talks continued without a break over lunch and dinner, and although I was exhausted when I got back to my hotel late that night, my head was buzzing and I found it hard to sleep.

What Wehner revealed to me was quite humbling: a small insect capable of performing navigational feats that we humans can only manage with the help of instruments. But I could not help being impressed by something else besides: the ingenuity and dedication of the scientists who had made all these discoveries.

Wehner was born in Bavaria in 1940, though his first memories are of being dug from the rubble of Dresden after the British bombing raid and subsequent firestorm that almost obliterated the city. During his primary school days he lived just outside the city in a house surrounded by a large garden, and it was in this "gorgeous bucolic setting" that he first developed an interest in natural history.

Later the family moved to West Germany, where he and his school friends spent their spare time studying songbirds—"counting clutch sizes, breeding times, feeding behavior, and the arrival and departure times of migrants." Though his father was a philologist and one grandfather a professor of languages, the young Wehner was strongly drawn to the natural sciences and, in 1960, he enrolled at the University of Frankfurt. There he attended courses in zoology, botany, and chemistry, and his interest "shifted more and more from the field to the lab, from natural history to physiology, especially biochemistry and neurophysiology." At this point, however, he had no idea that insects would become the focus of his work.

Scientists—at least, the best of them—spend as much time nurturing new talent as in pursuing their own research. Von Frisch certainly attracted and guided many excellent students who went on to do important research of their own. One of those who built on von Frisch's findings was Martin Lindauer, and in turn, he took the young Rüdiger Wehner under his wing.

In 1963, Martin Lindauer became director of the Institute of Zoology in Frankfurt, and his work on the sensory abilities of honeybees caught Wehner's attention. The prospect of carrying out rigorous experiments on freely moving animals fascinated him. From this point onward Wehner's ambition was to understand all the mechanisms that generated behavior: a causal chain that would lead all the way from the sensory organs to the brain cells that actually initiate movement. Wehner studied for his doctorate under Lindauer's supervision, exploring how honeybees distinguish different patterns, and then went to work at the University of Zürich, where he has been based ever since.[1]

As we sat together on that early summer's day, looking down on the calm waters of the lake, Wehner told me how, some months after receiving his doctorate, Lindauer had taken him to meet von Frisch at his famous Brunnwinkl estate in Austria. It was a largely symbolic occasion, and listening to Wehner's account, I was reminded of the "laying on of hands" that marks the apostolic succession in the Christian church.

The old master, though a wonderfully ingenious designer of experiments, was not at all comfortable with modern statistical methods. At the end of the interview, von Frisch, with a poker face, asked the young researcher, "I wonder, Dr. Wehner, how many legs does an insect have?"

This was, to say the least, a surprising question. Taken completely off guard, Wehner said hesitantly that most people assumed it was six. To which von Frisch responded with a smile, "I wouldn't be so sure nowadays. I would say 5.9, plus or minus 0.2!" Though this conversation occurred around the time when his work was under such fierce attack in the US, von Frisch seems to have preserved a dry sense of humor.

As a young post-doctoral student Wehner planned to follow in von Frisch's footsteps as a student of the honeybee but, as so often happens, the path of his career was altered by chance. Planning to perform some experiments in the spring, when honeybees are not yet on the wing in Europe, he drove out to Ramla in Israel, where he set up his apparatus in the middle of an orange grove. This was not a good site to choose. The trees were boiling with blossom and, not surprisingly, his bees preferred to gorge on this ready supply of natural nectar rather than paying any attention to the sugar solution on offer from him.

A dispirited Wehner was wondering how to attract the bees, when some long-legged ants caught his attention. As he watched them cantering around he became more and more fascinated by their behavior, and started some pilot experiments on their navigational abilities. The results were promising, but at this time Wehner knew nothing about the animals he was studying, not even their scientific name: *Cataglyphis*.

Although he did not realize it, he had found his ideal experimental subject.

Returning to Zürich, Wehner announced that he wanted to work on *Cataglyphis*, alongside his existing projects with honeybees. His scientific mentors all advised him not to devote his time to the study of such a "peculiar organism." Wehner listened to their advice, but ignored it. That turned out to be a good decision, but before he could put it into effect he needed to raise some money. Wehner also had to find a place where the desert ant flourished, but closer to home than Israel. He sat down with an atlas and worked out that the nearest practical location was Tunisia, the very place where Santschi had lived and worked sixty years earlier, though Wehner was unaware of him at this point.

North African adventures

In 1969, accompanied by a couple of students, Wehner set off by road and ferry for North Africa. They traveled south to the Chott el

Djerid—a salt pan near the oasis of Gabès in southern Tunisia—and it was there that they first encountered a foraging desert ant, which they later identified as a member of the species *Cataglyphis fortis*. This long-legged insect was dashing around under the broiling sun in search of food and eventually seized the remains of a dead fly. Wehner was amazed to see that it then ran straight back to its nest, no more than a small hole in the ground, which was more than 100 meters away. Since it could not possibly have seen the entrance from that distance, how was that possible?

For six weeks they worked in the desert near Gabès, but curious passersby interrupted them so often that Wehner decided to look for a more remote site. Later in the year, he returned to Tunisia with a small team of students. They soon found the ideal place: some salty sand flats near Maharès, a coastal town—then little more than a village—and set up camp. At that time, Wehner had no idea that this expedition would mark the start of a scientific career dedicated largely to the desert ant, or that he would be coming back to Tunisia every summer for more than thirty years.

Maharès in 1968 was no holiday destination, but Wehner and his wife Sibylle, who is also a biologist and has worked alongside him on nearly all his desert expeditions, were tough and resourceful. Food was not easy to come by, and their work exposed them to the exhausting heat of the desert. With the help of a local administrator, they found some simple accommodation on the upper floor of a local man's house, but their activities provoked a good deal of puzzlement among the villagers, and sometimes even suspicion. On one occasion, the local police mistook the Wehners for spies, and it was only Sibylle's linguistic skills that saved them from getting into serious trouble.

Santschi had long before shown that desert ants could find their way home, even when all they could see of the sky was a narrow circle defined by a cardboard cylinder. Von Frisch had later found that honeybees had access to a kind of sun compass based on polarized light. It seemed likely that the ants were using the same system, though nobody knew. And exactly how such a system actually worked—even in bees—was a mystery. Here then was a worthy challenge.

Wehner decided first to explore what role the ant's eyes played in accomplishing its navigational task. Ants, of course, are much easier to follow than bees, and Wehner was soon tracking them across the baking sands using a cunningly contrived wheeled frame, which held a variety of different filters above them as they ran around. This "rolling optical laboratory" also shielded the ants from the wind and prevented them from seeing any landmarks. With its help, Wehner established that their homing abilities did indeed depend partly on their sensitivity to polarized light.

Back in the laboratory, using an electron microscope, Wehner discovered a patch of cells along the skyward-facing (dorsal) edge of the ant's eye, which seemed perfectly designed to respond to light of this kind. By painting over different parts of the ants' tiny compound eyes, Wehner was able to show that this so-called dorsal rim area (DRA) was not only the key to the ant's ability to detect polarized light, but also supported a time-compensated sun compass. This discovery, which was soon extended to the honeybee, was a major breakthrough. Almost every insect since examined has turned out to have a similar specialized region for detecting polarized light. The DRA is in fact the basis of a standard insect compass, the evolutionary origins of which must be very remote in time.

Wehner wanted next to find out which parts of the ant's brain processed the signals from the DRA, but it is so tiny—smaller than a small pinhead—that it was impossible to investigate the behavior of individual cells within it. Instead, he and his colleagues had to rely on analogies drawn from work on the much larger brains of crickets and locusts to get a sense of the processes on which the ant's polarized light compass depends. They soon identified brain cells that responded to polarized light, and much of the circuitry involved in the processing of polarized light information has since been unraveled.[2]

The ant is certainly not a miniature version of a human celestial navigator. It does not perform complex calculations to compensate for the sun's movement across the sky. It does not need to, because it has a much simpler system at its disposal.

There are two parts to it. First, the desert ant employs what Wehner has described, using an idea drawn from engineering, as a "matched filter."[3] The ant literally matches what it sees with a model of the e-vector patterns in the sky that is built into its eyes. This physical template automatically determines the direction of the sun and the ant sets its course accordingly.

Then, just as in the case of the honeybee, a second mechanism comes into play. This is an internal clock ticking away inside the ant's brain, which allows it to compensate for the sun's changing azimuth. Under normal conditions this works well, though the ants may go astray if they cannot see the whole polarization pattern—for example, when clouds cover part of the sky.

The foraging desert ant, just like one of Bagnold's LRDG navigators, relies on a sun compass to maintain a steady course in the featureless environment of the desert salt pans. But a compass alone would not enable it to find its way home; DR also calls for a method of measuring distance. How on earth is an ant supposed to do that?

One possibility is that the ant makes use of a visual effect that scientists describe as "optic flow." Though it sounds complex, this is quite a simple concept: As we move, the scenery around us appears to flow past us at a rate that depends partly on its distance from us and partly on how fast we are going. As we look to either side, objects that are near to us seem to move faster than those that are farther away, while those that are directly in front of us seem to get bigger as we approach them. Ingenious experiments have shown how honeybees make use of this "flow" both to avoid obstacles and make safe landings, as well as to keep track of how far they have traveled on their foraging trips.[4] Optic flow "measurements" are one of the factors that shape the dances they perform on the surface of the hive.

Desert ants also make use of optic flow to work out how far they have gone on their foraging trips, but it turns out not to be the most important factor. Something else is going on.

The ant odometer

As long ago as 1904 it was suggested that ants might be able to measure distance by counting their steps, in the same way as the LRDG navigators relied on the odometers in their trucks (which count revolutions of their wheels) to keep track of how much ground they had covered. This was an intriguing theory, but nobody had found a way of testing it until Wehner's student, Matthias Wittlinger, had the bright idea of physically altering the length of the ant's strides—and found a practical, if drastic, way of doing so.[5]

First, Wittlinger trained normal ants to walk to and from a feeder ten meters away from their nest. Next, he transferred them to a high-sided test channel, in the same location, that prevented them from seeing any landmarks that would give away the position of their nest. Having deposited them at the feeder end of the channel, he measured how far they traveled homeward before they started to look around for their nest. These trained ants then underwent what is euphemistically known as an "experimental manipulation."

Wittlinger either attached stilts made of pig bristle to their legs (thereby lengthening their strides), or cut their legs off short (with the opposite effect)—a Draconian procedure the ants apparently tolerated with surprising equanimity. Both the stilt walkers and the amputees were then released at the far end of the test channel. He wanted to see whether the alteration in leg length affected how far they walked before starting to search for their nest. The results were dramatic: Those with stilts overshot the nest location, while the stump-walkers by contrast stopped well short of it. As predicted by the theory, it appeared that the stilt-walkers were overestimating the distance back to the nest, while the amputees were making the opposite mistake.

But that was not all. Wittlinger next allowed the manipulated ants to make the outward journey under their own steam, either with lengthened or shortened strides. They then behaved almost exactly like normal ants and correctly estimated where the nest lay. This made sense because, whether their legs had been lengthened or shortened, the same number of steps was required to make the outward journey as to return home.

With the help of its sun compass and odometer, the desert ant can find its way straight back to its starting point, the nest. What is more, it can do so no matter how tortuous its outward journey may be. It is a perfect example of DR in action. However, like human DR, the ant's system is not perfect. It is prone to cumulative errors, and as *Cataglyphis* may travel several hundred meters from its nest, these errors can become large.

In order to find out how the ants deal with this problem, Wehner placed a couple of black cylinders at the same distance on opposite sides of the ants' nest. The ants soon learned to use these prominent landmarks to locate their home. But it was not clear what features of the cylinders the ants were paying attention to. They might have been judging the location of the nest by measuring how far it was from the two cylinders, or they might have been working out what compass courses linked the cylinders to the nest—a form of triangulation. So Wehner and his colleague transported the ants to a test area well away from their real home, and set up the same array, but with some differences.

When the researchers doubled the *distance* between the cylinders (without altering their size) the ants did not, as you might expect, search halfway between them. Instead, they fussed around either one or the other of them. But when the *size* of the cylinders was also doubled, the ants behaved quite differently; now they were attracted to the midpoint.

Wehner concluded that they were seeking out a position from which the two cylinders *looked* just as they did during the original training session. Searching for their home, the displaced ants were trying to match a two-dimensional "snapshot" of the original array with what they saw now. So they trotted around until they could make the best match between the learned "template" and the current image of the cylinders detected by their compound eyes.

Warrant's sweat bees, you will recall, turn back and look at their nest from different directions when setting off on their journeys. Desert ants do something very similar. They perform "learning walks," during which they circle around their nest in ever-widening loops. Every now and then, they stop briefly and gaze back at the almost invisible entrance. While doing so, they memorize the views from various vantage points.

On returning from a foraging expedition they pull up these images and use them to find their way home. This image-matching system does not call for the ant to understand the geometrical relationships between the landmarks. In this respect it differs from the honeybee, which, remarkably enough, can learn how a group of landmarks relates to a food source in terms of the compass bearings connecting them, just like Clark's nutcracker.[6]

Building on these findings, Wehner and his colleagues have even managed to program a robotic vehicle that replicates the ant's polarized sunlight compass and landmark-recognition system. Playfully called the "Sahabot" (short for "Sahara Robot"), it can perform just the same maneuvers as a real ant.[7] They have also revealed many other aspects of the desert ant's navigational toolkit, including its ability to use the direction of the wind, vibrations, and scent as additional compass cues to find a target. The ants can even make allowance for the undulating surface of the ground in judging how far they have traveled. And the latest news is that these remarkable animals can also orient themselves using the earth's magnetic field.[8] There seems to be no end to their talents.

The desert ant lives in an extremely harsh environment and often faces temperatures so high that it can only endure being outside for brief periods. For this reason, it has long legs that keep it well clear of the hot ground, as well as enabling it to run very fast; Wehner has aptly described it as "the racehorse of the insect world." One species even has specially shaped hairs on its body that help it to control its body temperature.[9] Its ability to find the shortest route back to the shelter of its nest is more than a matter of efficiency; its very life depends on it.

Darwin was deeply impressed by the "wonderfully diversified instincts, mental powers, and affections" of ants and described the ant's central nervous system as "one of the most marvelous atoms of matter in the world, perhaps more so than the brain of a man."[10] He would surely have been delighted—and fascinated—to learn of Wehner's discoveries.

According to Stanley Heinze, a neuroscientist who studies insect navigation at the University of Lund: "One of the main functions of all brains is to take sensory information, use it to generate an estimate of the current state of the world, and then to compare it to the desired state of the world. If the two do not match, compensatory action is initiated, which is what we call behaviour."[11] That is just as true of insects as it is of more complicated animals, like humans.

Compared with birds and mammals, insects have tiny brains. While the human brain contains something like 85 billion neurons, that of a desert ant only runs to around 400,000. But even though their brains are small, and far less versatile than our own, they are perfectly adapted to the limited range of tasks they have to perform. While most of their behavior is controlled by brain circuits that are "hardwired," ants and bees (and other insects) can—as we have seen—learn from experience and generate an impressively diverse repertoire of navigational behavior. It is no wonder that designers of robots and autonomous vehicles look to them for ideas.[12]

The brains of insects as various as desert ants, fruit flies, moths, bees, locusts, and cockroaches contain two structures that seem to be of great navigational importance. The so-called mushroom body stores long-term memories based on smell and sight, while the "central complex" controls the course that the animal follows, in many cases making use of skylight polarization patterns to do so. Because these structures are shared so widely, it is thought that they must have emerged at a very early stage in the evolutionary process. Exactly how the animal chooses which way to go and initiates the appropriate movements is still mysterious, but interactions between the mushroom body and the central complex seem to play a crucial part in the process.[13]

∾

The estuarine crocodiles of Southeast Asia and Australasia are the world's largest reptiles—and have the unpopular habit of eating unwary humans. They may give the appearance of being quite sedentary, but they can move fast over short distances and can travel hundreds of miles at a more modest pace.

In 2007, a fascinating study revealed that they are also remarkably good at finding their way home. Three adult males were captured and fitted with satellite trackers. They were then carried in slings under a helicopter to different release sites on the Cape York Peninsula in Queensland, Australia. After spending some time apparently thinking about what to do next, all three of them eventually headed off and returned to the exact places where they had been captured.

One of the crocodiles traveled 62 miles along the coast in fifteen days; another covered 32 miles in only five days. That was quite impressive, but nothing compared to what the third one did. It was transported right across the Cape York Peninsula from west to east—an overland distance of 78 miles. Obviously it could not retrace its journey, but it still managed to get home by paddling right around the northern end of the peninsula and down the other side. It covered a distance of 255 miles in just twenty days.

Nobody has any idea how these animals found their way home, but this experiment provided a valuable practical lesson: There is clearly little point in "translocating" crocodiles that pose a threat to people.[14]

Chapter 8

STEERING BY THE SHAPE OF THE SKY

More than half of the human race is cut off from the most sublime spectacle that nature has to offer. Living in cities and towns where the night skies are aglow with man-made light, most of us can see only a handful of the thousands of stars visible in places beyond the reach of light pollution. We have slowly but surely been drawing the blinds on a window that once offered us a view of the universe.

When the power supply to Los Angeles was knocked out by an earthquake in 1994, the sight of a truly dark night sky was so unfamiliar that many residents called the emergency services, anxiously reporting a strange "giant, silvery cloud" in the sky. Were aliens about to land? No, but it *was* something they had never seen before: the Milky Way.[1]

According to recent research[2] based on satellite images, more than 80 percent of the world and more than 99 percent of the US and European populations live under light-polluted skies. The Milky Way is hidden from more than one-third of humanity, including 60 percent of Europeans and nearly 80 percent of North Americans. The scourge of light pollution has crept up on us so slowly that hardly anyone realizes how much it has cost us—and it is getting steadily worse.[3] It is damaging to human health,[4] and other animals that depend on natural light for many different purposes, including navigation, are suffering even more from its ill effects.[5] Many die as a direct result of the disruptive effects of artificial light on their normal routines. It is a serious environmental problem that receives far too little attention.[6]

To witness the velvet blackness of a sky filled with stars, you have to travel into the desert or mountains, or far out onto the open sea. If you have the chance to visit one of these remote places on a clear night, you will discover how the heavens must have once looked to everyone.

At first you can see only the brightest of the stars, but as your eyes slowly adapt, more and more come out, until at last the sky is alive with thousands of glittering points of light. And then you start to see how the stars differ from each other, not just in brightness, but also in hue. Some have a reddish, some a yellow tint, while others—the hottest ones—gleam with an icy, blue-white light. Though we can see only our nearest heavenly neighbors with the naked eye, even they are unimaginably distant from us; the star Deneb, for example, is more than a thousand light years away. Since light travels about 186,000 miles each second, that is a very long way.

It was on the open ocean that I first saw a sky like that—and it was a revelation. Although I had long been fascinated by the stars, I had never realized what an overwhelming sight they could be in all their glory. Then, as I kept watch hour after hour, I saw for the first time how they moved.

Around the still point of Polaris, the whole night sky with all its stars was majestically turning—in time with the slow rotation of the earth. And sitting there aboard a small yacht in the middle of a big ocean, looking out into the depths of space, I was crushingly aware of my own insignificance. Yet, oddly enough, that feeling was not at all troubling. It was in fact strangely calming.

People have been looking up at the stars for a very long time: 300,000 years or so, if the latest estimates of the age of *Homo sapiens* are to be relied on. And our earliest ancestors must have regarded the night sky with at least as much wonder as any of us alive today. They must also have realized that it exhibited certain regularities that could be of use to them, and it would be strange indeed if other animals had not begun to exploit them at a much earlier date.

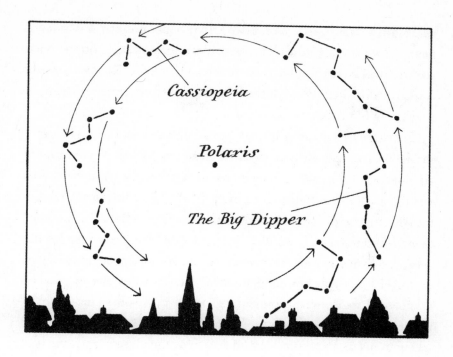

Cassiopeia

Polaris

The Big Dipper

Leaving aside how the different constellations seem to come and go according to the changing seasons, the earliest humans would have noticed that each star follows a regular daily course, just like the sun. Apart from those that lie close to the celestial poles—the points in the sky vertically above each of the geographical poles—they all rise in the east and set in the west. And, like the sun, they always bear due north or south of the observer when they reach the top of their arcs, or cross your meridian. Though Polaris has not always marked the northern celestial pole (as it does today), prehistoric astronomers would surely have noticed that there was a still point amid the stars that circle both the northern and southern celestial poles.[7]

Our Stone-Age ancestors must have watched the skies very closely. They were well aware of celestial events like the summer and winter solstices, and they built many structures (of which Stonehenge is one of the most famous) that are carefully aligned with them. Later on, the remarkably sophisticated observations of the Babylonians, Greeks, and Arabs provided the foundations on which modern astronomy was built. We also know that ancient European, near-Eastern, and Chinese seafarers made long voyages on the open sea and well beyond sight of land. They must have used the sun and stars to guide them, though the historic record sheds very little light on exactly how they did so.

There are tantalizing glimpses, such as the lines in Homer's *Odyssey* in which Circe tells the hero to steer a steady easterly course by keeping the stars of the Great Bear always on his left, but the earliest detailed written accounts of the practice of navigation date only from the six-teenth century. Before that date, we are largely in the dark. Literacy being limited to a very small, privileged élite, the craft of navigation was presumably passed on by word-of-mouth and practical example.

Nevertheless, the few indigenous societies that have not succumbed completely to Western domination can give us some clues. By the mid-twentieth century, it was only in a few remote places that ancient navigational methods survived, and of these, the most famous and best studied are the traditional techniques employed by the islanders of the Pacific Ocean.

The European sailors who first reached the Pacific during the sixteenth century were astonished by the navigational skills of the peoples they encountered, though they had great difficulty in under-standing them. It was not until the first scientific investigators arrived in the second half of the eighteenth century that brief descriptions of the navigational methods of the Polynesians began to appear in print.

Louis-Antoine de Bougainville (1729–1811), the great French explorer whose arrival in Tahiti in 1768 shortly preceded that of Cook, was amazed to find that the islanders were able to make suc-cessful landfalls on distant islands after crossing hundreds or even thousands of miles of open ocean, without making use of instruments or charts of any kind. And Cook himself was so impressed by the knowledge and skills of one Tahitian navigator that he took him aboard his ship to help in exploring the neighboring islands—and eventually also New Zealand.

But the descriptions of Polynesian navigation by Bougainville, Cook, and their companions are frustratingly skimpy. Perhaps they asked the wrong questions, or maybe the islanders were reluctant to share such vital—indeed sacred—information with their guests. Quite apart from language problems, the radical conceptual differences between the European and Polynesian approaches to navigation may well have obstructed successful communication. In any event, over the next two centuries, the crushing impact of colonial power and influence came close to smothering the techniques that had, over a period of several thousand years, enabled the Polynesian peoples not only to populate but also to maintain regular contact between islands flung across half the Pacific Ocean. The Western investigators in the 1960s, who began to seek out the few surviving practitioners of the ancient skills, very nearly arrived too late.

Star courses

By that time traditional navigation had been abandoned in the islands of Polynesia, but it was hanging on in the Micronesian archipelago. Sailors there were still making voyages that took them across hundreds of miles of open sea using age-old methods. The key to their success lay in their long apprenticeships, which might begin before the age of ten, during which they would learn through endless reiteration and testing the "star courses" that linked all the islands to which they might need to sail.[7]

These courses were shaped by knowing exactly where on the horizon thirty-two named stars rose and set. This "star compass" system was so deeply ingrained and thoroughly understood that the navigator could set an accurate course, not only when a familiar star was dead ahead, but also when it appeared at any other point in the sky (though obviously not when it was vertically overhead).[8] He—and it was always a man, as women were forbidden to be navigators—steered "by the shape of the sky," rather than aiming for a single point of light.[9]

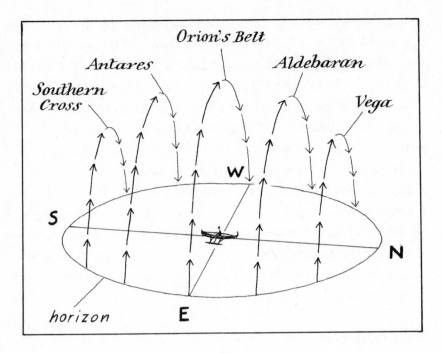

Southern Cross · Antares · Orion's Belt · Aldebaran · Vega

W

S

N

horizon

E

Some of the brightest stars that form
part of the "star compass" of the Pacific
Islanders

The star compass may have been the heart of the Micronesian navigational system, but it was by itself not enough to make long-distance navigation practical.[10] The helmsman also needed to be able to steer a course during the day, and that meant using the sun. In the tropics, the sun generally rises close to due east and sets close to due west. At noon, when it reaches its highest point in the sky, and provided it is not directly overhead, it also tells the navigator where north and south are.

For the rest of the day, the navigator must once again steer "by the shape of the sky." According to David Lewis, an extremely experienced and intrepid small-boat sailor, who was also an expert on traditional navigation in the Pacific Islands: "If one knows the sun's bearings at rise and set and its track across the sky, with sufficient practice it becomes habitual to make the almost automatic mental interpolations that are needed for steering by the sun."[11] As we have already seen, even ordinary people can make surprisingly good use of the sun and moon to maintain a steady course if the need arises.

But celestial navigation was not enough. The master navigator needed also to be an expert in the art of DR. He had to be able to judge the speed of his voyaging canoe with great accuracy and to take account of the often powerful effects of ocean currents. Changes in the color of the water and in the shape of the waves enabled him to detect the presence of submerged reefs. These helped him check his position along the route, even when out of sight of land.[12]

Local, wind-driven waves are fickle and may head in any direction, and are therefore of little navigational value on the open sea, but the regular swell generated by distant weather systems is much more useful. Rolling majestically along, it can easily travel for hundreds or even thousands of miles, and it always heads in the same direction until it encounters land. To the navigator, such swells functioned like a compass, enabling him to maintain a straight course even when the sky was completely obscured by clouds.[13]

In some parts of the Pacific, navigators were able to detect the presence of a still invisible island by the ways in which it disturbed the regular swells around it.[14] In the Marshall Islands, special "charts" were made out of sticks to illustrate the characteristic patterns produced by the reflection and diffraction of swells around an oceanic island. Though not used at sea, these seem to have served as useful teaching aids.

Clouds clinging to the slopes of high islands acted as valuable long-range beacons. They would also reflect the characteristic pale green light from the shallow lagoon within a distant atoll. But the main method of locating an as-yet invisible target island was to observe the course taken by birds returning to their roosts at sunset. Since land-based birds often travel far out to sea to look for food, they can reveal to the knowledgeable navigator the presence of land as much as forty or fifty miles away.[15]

The many and varied skills employed by traditional Pacific Island navigators have been revived in recent years. The Polynesian Voyaging Society, based in Hawaii, has been at the forefront of this process and under its aegis replicas of traditional long-range voyaging canoes have undertaken some remarkable journeys using the old navigational techniques. One such canoe, named *Hōkūle'a*—"star of gladness": the Hawaiian name for the star known in the West as Arcturus—completed a three-year, around-the-world voyage in 2017.

Apart from the oceans, the most imposing obstacles confronting migrating animals are the world's great mountain ranges, but some animals are equal even to that challenge.

A mountaineer on the 1953 Everest expedition named George Lowe, who was also an expert ornithologist, claimed to have spotted bar-headed geese flying above the summit as he sat on the high slopes of the mountain. And the naturalist Lawrence Swan later described hearing geese fly overhead as he stood on the Barun Glacier beneath the 8,485-meter Himalayan peak of Mount Makalu on a cold, still night: "Coming from the south, the distant hum became a call. Then, as if from the stars above me, I heard the honking of bar-headed geese."[16]

Unlike human mountaineers, who need to acclimatize themselves before attempting to climb at high altitudes, bar-headed geese can apparently cope with the extremely thin air by massively increasing the rate at which their hearts beat,[17] though they normally follow the valleys when crossing the Himalayas, rather than going over the summits. The great mountain chain did not even exist when the ancestors of today's geese first began making their migratory journey. As the land started to rise—about twenty million years ago—it is thought that successive generations of geese gradually adapted to the growing demands placed upon them.

Chapter 9

HOW BIRDS FIND
TRUE NORTH

The shrieks of the common swifts zooming past my window, as they hunt their insect prey, seem to be filled with a fierce joy. I welcome their arrival as the first proof that the summer has finally arrived. These wonderfully quick and agile flyers rarely land, except to build their muddy nests and feed their young; we now know that, outside the breeding season, they can remain aloft for as long as ten months. Provided they can find enough food and water along the way, a journey from Africa to northern Europe presents them with no trouble at all, though whether they sleep on the wing—as some frigate birds apparently do—remains a mystery.[1]

The seasonal comings and goings of birds baffled the ancients, and they came up with some strange explanations for what they observed. Aristotle (384–322 BCE) believed that the redstarts he saw in the summer were the very same birds as the robins that appeared in winter, but transmuted; they did not go anywhere, they simply changed their spots, so to speak.[2] In fact, as we have since learned, these separate species were migrating in opposite directions—and exchanging places.

In 1555, a woodcut of a man netting swallows from a lake appeared in a book by the Swedish archbishop, Olaus Magnus. He claimed that this was where they passed the winter months, and that birds caught in this way could be revived by warmth, though they did not long survive.[3] And as late as 1703, an Englishman by the name of Charles

Morton wrote a pamphlet arguing, apparently in all seriousness, that storks passed the winter months on the moon.[4]

The Reverend Gilbert White (1720–93), a clergyman living in the small English village of Selborne, was puzzled by the phenomenon of migration, but he was in no doubt that it was real. Writing in 1771 to a skeptical correspondent, who was "no great friend to migration," he insisted that even if some swallows passed the winter months in slumber:

> . . . migration certainly does subsist in some places, as my brother in Andalusia [the southernmost province of Spain] has fully informed me. Of the motions of these birds he has ocular demonstration for many weeks together, both spring and fall: during which periods myriads of the swallow kind traverse the Straits [of Gibraltar] from north to south, and from south to north, according to the season.[5]

Perhaps the first really hard evidence of long-distance bird migration—all too hard from the point of view of the bird in question—was the discovery in a north German village of a still-living stork, transfixed by an arrow of indubitably African origin. That was in 1822, and the stuffed body of that stork found its way into the collection of the Zoological Museum in Rostock, where it can still be seen today. The appearance of this so-called *pfeilstorch* ("arrow-stork"), and subsequently quite a few more hardy survivors of African archery, proved that some birds did indeed travel immense distances during their annual migrations.[6]

The legendary American ornithologist and artist John James Audubon (1785–1851) added another piece to the puzzle. His magnificent set of prints entitled *The Birds of America* appeared in the 1830s, and in the accompanying text he described attaching fine silver threads to the legs of young pee-wee flycatchers at their nest near his home in Pennsylvania. He observed that after heading south in the autumn, the very same birds, identifiable by their silver adornments, returned in the spring to their place of birth. This piece of evidence showed that at least some migratory birds returned faithfully to the same nesting sites year after year.[7]

The Danish schoolteacher and ornithologist Hans Christian Mortensen (1856–1921) carried out the first successful experiments using bird banding in 1899. Rather than using silver threads, he employed aluminum metal tags that carried an identification code and a return address. This technique has since played a crucial part in establishing the patterns of migration in many bird species. Trapping and netting have also proved helpful, especially at migratory hotspots where birds pass in vast numbers, such as Rybachy on the Baltic coast of Russia.

But the electronic revolution has completely transformed our knowledge of animal navigation. Since its development during the Second World War, radar has been widely used to monitor migratory birds and is also used to track flying insects such as bees and moths. In addition to various kinds of "data-loggers," which record information that can be downloaded later, tracking devices with inbuilt GPS chips signal an animal's precise position to overhead satellites in real time—and miniaturization means that these tools can now be deployed even on quite small birds.

We have now entered the "golden age of animal tracking," and we can expect many discoveries that will shed light not merely on navigational behavior, but on a whole range of important environmental and ecological issues.[8]

Roughly half of all bird species migrate, and there is a wealth of data about the journeys they make. Some cover enormous distances—the arctic tern being only an extreme example. The North American bobolink flies from its breeding grounds in Canada all the way to Uruguay. Swainson's hawk follows a similar route, traveling in large flocks from the prairies of North America to the pampas of Argentina. Brent (or brant) geese breed in the high Arctic—farther north than any other geese—and some travel all the way from Wrangel Island, off the northeastern coast of Siberia, to Mexico, a journey that involves a nonstop passage over the Pacific of roughly 3,000 miles.

Birds of prey tend to avoid flying over water, but there is one spectacular exception. The amur falcon, a small insect-eating species that breeds during the summer months in Mongolia, Siberia, and northern China, flies about 8,000 miles to reach southern Africa toward the end of the year. This journey includes an ocean crossing of around 2,500 miles between southwestern India and East Africa, the longest over-water journey of any raptor.[9] It is possible that they keep themselves going *en route* by picking off migratory dragonflies that are heading in the same direction (*see pages 136–37*).[10]

Many migratory birds travel in mixed groups of adults and juveniles. One great advantage of this system is that it allows the youngsters to learn the correct migratory route from their elders.[11] In principle, members of these species could rely entirely on learned, landmark information to find their way, as each generation passes on its expertise to its successor. But it is hard to see how birds that fly for long distances over the open ocean could rely on such a technique, and birds that fly solo obviously cannot do so.

Lonely young cuckoos

Not all young migratory birds have the benefit of an adult guide. By the time a young European cuckoo leaves the nest of its foster parents, its biological parents will already have headed south for their overwintering grounds in southern and central Africa. The juvenile bird therefore has to find its own way. In common with many other migratory species, cuckoos travel at night, partly because the air is cooler (overheating can be a serious problem for birds in flight), and partly to avoid the attention of predators. A young cuckoo that has never made the journey before plainly cannot be following a route it has learned, so what navigational techniques can it be using?

It has long been assumed that young cuckoos rely on an inherited guidance program that essentially points them in the right direction and tells them to fly for a certain length of time. The theory is that such a "clock and compass" system would enable them to reach at least roughly the right area, but it does not square well with the findings of a recent tracking experiment.

This revealed that the young cuckoos follow a surprisingly narrow "corridor" and stop off along the way at the same staging areas to rest and refuel. After traveling more than 3,000 miles, it was found that the average distance between individual birds was a mere 102 miles.[12] These observations suggest that other factors are probably involved, including perhaps some inherited ability to recognize major landscape features which mark the correct route.

The extraordinary navigational skills of the young cuckoo remain mysterious, but these birds—in common with other solitary migrants, as well as those that make extended journeys over the featureless open ocean—must at least have access to some kind of compass that enables them to set and follow a steady heading.

As we know from insects, one possibility is that such a compass might be based on celestial cues—in other words, the patterns that can be observed in the sky.

Birds in the northern hemisphere could make use of Polaris. Its azimuth is always due north (true, not magnetic—the magnetic pole moves around and currently stands about 300 miles away from the geographic pole). That means that if you are looking directly at it, you must be heading north, if it is on your right you must be going west, and so on. A bird could therefore maintain a steady course, in any direction, simply by making sure that the relative bearing of Polaris remained unaltered. There would be no need to employ any kind of clock or to perform any calculations. One of the most widely used tools for studying the migratory behavior of birds is the Emlen funnel. Invented by Stephen Emlen this almost absurdly simple device takes advantage of the fact that captive birds will repeatedly try to escape from a cage in their preferred migratory direction. In the traditional Emlen funnel, the bird stands on an ink pad at the narrow end of the funnel and, as it jumps up and down trying to fly off its feet leave inky marks on the paper lining its sloping sides. The resultant graffiti are presumed to show where it wants to go.

In the late 1950s, Franz Sauer had the bright idea of testing how birds reacted to the patterns of stars presented in a planetarium and concluded, on the basis of an admittedly small sample, that they were able to make good use of them for navigational purposes.[13] Later work by Emlen, employing his famous funnels, showed that indigo buntings, while not apparently paying attention to any individual star, were able to detect the revolving pattern of stars around Polaris.[14]

The buntings were able to identify the centre of this rotational pattern, though the exact arrangement of stars within it did not bother them. When the birds were shown a night sky that revolved around Betelgeuse—a bright star in the constellation Orion—rather than Polaris, they were completely unfazed and set their course accordingly.[15] The fact that they become disoriented when their view of the stars is obscured shows how important the patterns are to them. So it is easy to see why light pollution poses such a threat to them: Navigating accurately by the stars is only possible if they are visible.

It seems that many other birds that migrate by night find true north in the same way.[16] The great advantage of such a system is that, once learned, it is simple to use and, unlike the sun compass, requires no form of time-compensation. Yet it is still not clear how birds learn to recognize these patterns in the night sky. It is hard to believe that they can perceive the motion of the stars, as it is so slow, but they may be able to infer it by comparing "snapshots" of the sky at intervals during the night.

As long ago as the 1930s, the ornithologist and writer Ronald Lockley, who lived on the small island of Skokholm, off the southwest coast of Wales, showed that Manx shearwaters were capable of astounding feats of long-distance navigation. He took two of these wild seabirds from Skokholm by air to Venice—a place they would never normally be expected to visit. One of them nevertheless found its way back to its burrow in just two weeks.

But that was nothing compared to what happened in 1953, when Lockley persuaded a visiting musician called Rosario Mazzeo to take a couple of shearwaters with him on his return to the US:

> I left Tenby, Pembrokeshire, that evening via sleeper train for London. The birds caused no little wonder and merriment to the people in the adjoining rooms, who could not understand the origin of the mewing and cackling sounds which came from my room in the late evening. The next day the birds remained in the carton, each in its own compartment, and in the evening, I enplaned for America with the birds under my seat.

Sadly, only one of the birds survived what must have been a very stressful journey, and Mazzeo immediately released it on arrival in Boston. The distance from there to Skokholm is more than 3,000 miles, but the bird (which had been banded) got home to its burrow after only 12.5 days—in fact, it arrived before the letter reporting its release. Understandably the person who found it was "completely flabbergasted."[17]

Chapter 10

HEAVENLY DUNG BEETLES

For an idle half hour in the south of France, I was entranced by a shiny black dung beetle as it tried—repeatedly and indefatigably—to roll its ball up and over a small but steep ridge. Time after time it lost control as it neared the top, and had to go back down and start all over again, but eventually it succeeded and I felt like applauding.

The ancient Egyptians worshipped the dung beetle, believing that it symbolized the sun god, Khepri, who rolled the ball of the sun across the sky. Eric Warrant, who has worked with them over many years, admires dung beetles almost as much: "They are so determined. That's what makes them so wonderful to work with. In many ways they're like little machines: they'll roll balls forever, at any time."

Rolling a ball in a straight line may not sound like a very impressive feat. But bear in mind that the beetle has first got to sculpt the dung into an accurate sphere (or it will not roll at all); it then has to go backward steering the ball with its hindmost pair of legs, across ground that may be very uneven.

Over the last twenty years, Warrant and his colleague, Marie Dacke, have conducted a succession of fascinating experiments on dung beetle navigation that have attracted a good deal of public attention, not least in the form of an Ig Nobel Prize. These are awarded in Boston every year for scientific research that "first makes you laugh, then makes you think." They are designed to draw attention to the sheer strangeness of the universe around us—and the extraordinary, often eccentric, dedication of the scientists who investigate it.

Although they are not meant to be taken too seriously, they are in their own way very prestigious, and real Nobel Prize winners always attend the ceremony. When Warrant and his team received their prize, a small girl stood on the stage while each winner made a short speech describing their work to the large audience. The girl's job was to tell the speaker to shut up when she thought they were getting boring. Warrant was one of the few who managed to get through their speech uninterrupted.

At the start of his scientific career, Warrant was studying how dung beetles see in the dark. African dung beetles (also known as scarabs) were introduced to Australia to deal with a problem caused by an earlier animal import: the cow. The native dung beetles were only used to tackling kangaroo droppings, and they had no idea what to do with the accumulating piles of cow dung that were causing serious agricultural damage. For the newly arrived African scarabs, Australia must have been like heaven—vast accumulations of dung and no competition. They quickly and efficiently started burying all the stuff their local cousins had been ignoring, and thereby restored the productivity of Australian pastures, without apparently causing any problems for other animals.

In 1996, Warrant attended a conference on dung beetle biology in South Africa's Kruger National Park. There he heard for the first time about *ball-rolling* dung beetles. Unlike the ones he was familiar with, these beetles scoop up the dung, deftly shape it into small spheres and then roll them away as fast as they can. They then eat the dung or lay eggs in the balls and bury them to provide food for their young when they hatch.

Warrant recalls hearing the speaker say: "It's amazing, they just roll their balls in straight lines all the time and I don't know how." He was sitting in the audience thinking excitedly, I know, I know: They must be using the polarized light patterns in the night sky! He put his hand up, asked a question, and the course of his career was changed.

Warrant and his colleagues soon showed that the ball-rolling scarab had a DRA for detecting polarized light, exactly like the desert ant. Then he and Marie Dacke began to explore how the beetles actually used it for navigational purposes. Evidently the competition for dung among these beetles is intense, and in order to make a quick getaway, the beetle must roll its ball away from the dung pile in as straight a line as possible. Otherwise it runs the risk of getting into a scuffle with other beetles and being robbed of its precious cargo. Before it sets off, the dung beetle climbs on top of its newly-formed ball and performs a curious circular dance during which it carefully inspects the sky above it.[1]

Many insects are nocturnal, but their compound eyes, though extremely sensitive in low light conditions, offer far less visual acuity than those of birds, or humans. So, while they can see much better than us in the dark, their visual world is far blurrier than ours. It is doubtful that the dung beetle can see many individual stars, except perhaps the very brightest ones.

The most obvious possibility is that it uses the brightest light source in the night sky: the moon. Since its forays are of short duration, the dung beetle has no need to make any allowance for the moon's changing azimuth, but the moon is nevertheless an inconstant guide. Its phase changes constantly, so the amount of sunlight it reflects varies greatly, and it rises and sets at a different time each day, too. To complicate matters further, there are several nights every (lunar) month when the "new" moon is so close in the sky to the sun that it cannot be seen at all. And the intensity of even the full moon's light is far, far lower than that of the sun, though its spectrum is much the same. It also includes ultraviolet light; in theory you could get moon-burn, though it would take a very long time.

The dung beetle is well adapted to cope with lunar vagaries. First of all, it relies not so much on the disk of the moon itself for guidance, as on the polarization patterns (e-vectors) of its light, in the same way that bees and desert ants use the polarized light of the sun during the day.[2]

Completely cloudy nights are not common in the part of South Africa where Warrant and Dacke conducted their experiments, but what is a beetle to do when there is no moon?

The discovery that beetles can set a course using polarized moonlight caused a big stir, and the paper describing it had the distinction of being published in the leading scientific journal, *Nature*. Some years later, however, Warrant and Dacke received a shock. They were in camp on a brilliantly clear night at the edge of the Kalahari Desert. The velvet-black sky was filled with stars and they were waiting for the moon to rise, so that they could start a new experiment.

Warrant described to me what happened next:

> We had some dung out to try to catch the beetles and they were flying in. Then they started making balls—the buggers!—and rolling them away in perfectly straight lines, with no polarized light . . . Both of us got very nervous, because suddenly it looked like "*Nature* retractions, *Nature* retractions!"

To be forced to withdraw an article from any scientific journal, because it has proven to be inaccurate, is a very public humiliation; but a retraction from a top one like *Nature* is about as bad as it can get. "A certain amount of drinking was going on at this point," according to Warrant, but eventually a thought struck the two of them:

> Just a minute, there's a big stripe of light across the sky! The Milky Way. Maybe they're using that—could they be using that? There's nothing else round here they could be using.

Putting hats on beetles

To test their new idea, Warrant and Dacke started by attaching little cardboard hats to the beetles, which made it impossible for them to see the sky. They then had much more trouble steering a straight course than when they had an unimpeded view. When transparent plastic caps were substituted for the cardboard ones, they managed perfectly well again, so it was plainly not the mere encumbrance of a cap that was putting them off their stride. The next step was to test the beetles in a circular arena, surrounded by a high barrier that prevented them from seeing any landmarks. The researchers also removed the overhead camera that recorded the beetles' movements in case this, too, because it was providing some kind of directional information.

They placed each beetle with a ball of dung in the middle of the arena and timed how long it took to reach the edge, which was marked by a circular chute. The clatter of the beetle crashing down into the chute told them when it had got there, and the length of time it had taken indicated the straightness of its path. Under these conditions, they were able to show that the beetles did indeed need to see the starry sky to maintain a straight course, though they performed even better when the moon was also present. However, under a cloudy sky they were disoriented.

The researchers now took the beetles and their arena to a planetarium. Under one condition, the animals could see the full starry sky, including a long streak of light imitating the Milky Way, but without the moon. Under another, they could see only the Milky Way. Their ball-rolling performance was not much worse when they could see the full starry sky, with the Milky Way, than when they could see the moon. And when the Milky Way alone was presented, they did almost equally well. However, when the long-suffering beetles had access to an array of 4,000 dim stars *without* the Milky Way, their performance deteriorated considerably; and when just eighteen guide stars were available, it got worse still.[3]

So, it appeared that the beetles were not making use of any individual guide stars. "This finding," as Dacke reported, "represents the first convincing demonstration of the use of the starry sky for orientation in insects and provides the first documented use of the Milky Way for orientation in the animal kingdom."

Although the individual guide stars were not of much help to the beetles, Warrant told me that it was still unclear whether the beetles could actually see them. He thinks they probably can, and this is something he hopes to be able to clarify by recording the responses of individual light-sensitive cells in the eye of the beetle, just as he has done with the sweat bee.

Dung beetles are not the only arthropods that can steer by the light of the moon. The large yellow underwing moth can apparently do so,[4] as can sandhoppers—small crustaceans that live a liminal, seaside existence. These animals, related to woodlice, are well named, because their natural escape reaction is to propel themselves wildly into the air by flexing their carapace. If you have ever built sandcastles, you may well have encountered them, though their numbers are in decline in many places.

It is not obvious why a creature as small and apparently primitive as a sandhopper should care about the position of the moon. The answer is that they are extremely fussy about moisture. They die if they dry out, but they drown if they are submerged in salt water. So they need constantly to move back and forth as the tides rise and fall, and they must also be able to find their way back to a nice patch of damp sand after their nocturnal foraging expeditions. And, of course, it is absolutely vital that they move in the right direction. The sandhopper is the Goldilocks of the arthropod world.

As long ago as the 1950s, two Italian scientists, Leo Pardi (1915–90) and Floriano Papi (1926–2016), made the extraordinary discovery that sandhoppers used both the sun and moon as compasses to help them move either toward or away from the sea, according to need. This ability apparently depends on two separate clocks, one calibrated to the daily movements of the sun and one to the slightly different lunar cycle.[5]

The sandhopper's sun compass is situated in its brain, while its lunar compass is based in its antennae. And whatever mechanisms govern these processes are plainly innate, because sandhoppers raised in captivity will always head off in the direction appropriate to their place of origin. This means that a sandhopper descended from ancestors on a south-facing coast will always tend to go south to find the sea, while another with ancestors on a north-facing coast will tend to go north.

There is at present evidence of only one animal—apart from Homo sapiens—*having the ability to navigate with the help of individual stars, as opposed to their circumpolar patterns, though it is not very strong. The animal in question is the harbor seal. A study involving only two animals, called Nick and Malte, was carried out in a custom-made swimming planetarium.*[6]

Both seals were taught to identify a "lodestar" (Sirius) from a projection of the night sky as seen in the northern hemisphere, and to indicate its position by swimming to the point on the edge of the pool immediately beneath it. Eventually, they were both able to perform this feat with some precision, reliably aiming within a degree or two of the azimuth of Sirius. On the strength of these findings, the authors argued that harbor seals might be able to develop a star compass system similar to that used by the Micronesian and Polynesian pilots:

> *We suggest that marine mammals might learn to identify lodestars in the pattern of the night sky and to use these lodestars as distant landmarks . . . to steer by in the open seas. This might be at least one possible mechanism of offshore orientation, until an expanded target like a coastal region is reached and goal-related terrestrial orientation mechanisms can correct their swimming direction.*

This is an appealing idea and, if correct, might help explain how many marine animals navigate, but—in the time-honored scientific phrase—more research is needed.[7]

Chapter 11

GIANT PEACOCKS

"Come quickly; come and see the moths, as big as birds!" With these words, Fabre's little son, Paul, rushed excitedly into his father's room. Huge, male giant peacock moths[1] seemed to have taken over almost the whole house. The maid was chasing them around madly, thinking they were bats. Holding a candle, Fabre went to the office, where he had left a freshly hatched female moth imprisoned under a muslin cheese cover earlier in the day:

> What we now see is unforgettable. Flapping gently, the great moths fly around the cover, settle, take off, come back, fly up to the ceiling, then come down from it again. They throw themselves on the candle, and put it out with a wingbeat; they crash down on our shoulders, cling to our cloths, brush against our faces. It is like a necromancer's lair with its whirling cloud of bats. For reassurance, little Paul grips my hand much harder than usual . . . Here, in effect, are forty lovers, who have come from all directions, alerted I know not how, determined to pay court to the marriageable female born that morning in the mysterious interior of my office.[2]

Fabre was left wondering what strange force had drawn so many amorous moths to his house through the warm Provençal night. He rightly suspected that a smell emitted by the female was the key, and that the elaborate, frilly antennae of the males might be the means by which they detected it. We now know that male moths like these can pick up the sexual scent released by a potential mate at a distance of several miles and follow it to its source. Many insects rely on scents to find their mates, to locate food, or to find suitable sites to lay their eggs.

Because the odor plumes the insects follow are diluted very rapidly as they disperse, they may initially be responding to a single scent molecule, but the plumes often break up completely in a moving airstream. So finding the source of an airborne smell is not, as was once believed, a simple matter of following a trail that steadily grows stronger (a "concentration gradient") to its source.[3]

Exactly how insects overcome this tough sensory challenge has been the subject of much debate.[4] In addition to casting from side to side to relocate the odor when they lose track of it, and heading generally upwind, Fabre's giant peacock moths were probably also making use of the different signals they received from each of their two extraordinarily sensitive antennae.

Honeybees certainly change course in response to differences in the chemical content of the air passing over each of their antennae,[5] as do fruit flies.[6] And recent experiments with the redoubtable desert ant have shown that it not only makes use of olfactory cues to find its nest (in addition to all the visual ones we discussed earlier), but also that it needs both of its antennae to do so efficiently.[7] The process of comparing the inputs from each antenna is colorfully described as "stereo olfaction," and it may even provide the animal with a kind of "odor compass."

In the late 1940s, as a young researcher, Arthur Hasler was trying to discover how fish use smell to tell the difference between plants. He was much impressed by the work of Konrad Lorenz, who had recently discovered the principle of "imprinting"—the rapid, irreversible form of learning that gives rise to rigidly fixed patterns of behavior in some animal species. Lorenz famously showed how newly hatched geese imprint on the first moving thing they see and will follow it blindly, even if it happens to be a scientist in Wellington boots, rather than mother goose herself.

Hasler was also fascinated by the way in which adult salmon, having spent several years eating, growing, and maturing out on the open ocean, returned to breed in the very streams where they had been born. This had been well established by tagging young fish and recapturing them. But how they managed this extraordinary feat was still a complete mystery.

While hiking in the remote Wasatch Mountains of Utah, Hasler underwent a revelatory experience:

> I had approached a waterfall, which was completely obstructed from view by a cliff; yet, when a cool breeze bearing the fragrance of mosses and columbine swept around the rocky abutment, the details of this waterfall and its setting on the face of the mountain suddenly leapt into my mind's eye. In fact, so impressive was this odor that it evoked a flood of memories of boyhood chums and deeds long since vanished from conscious memory.
>
> The association was so strong that I immediately applied it to the problem of salmon homing. The connection caused me to formulate the hypothesis that each stream contains a particular bouquet of fragrances to which salmon become imprinted before emigrating to the ocean, and which they subsequently use as a cue for identifying their natal tributary upon their return from the sea.[8]

Building on this insight, Hasler and his colleagues, in a succession of ingenious experiments, were able to demonstrate that salmon could in principle imprint on the very particular scents that characterize their natal streams, and use them to find their way home to them from the sea.

In the 1970s, Hasler successfully attracted salmon raised in hatcheries to rivers scented with either of two synthetic chemicals to which they had been briefly exposed years earlier. The fish could not have experienced either of these smells during the interval, but had nevertheless retained the memory of them. This very technique later proved useful in attracting salmon back to the cleaned-up Great Lakes—from which pollution had banished them in the past.[9]

That homing salmon rely on olfactory signals is now well established. But in the wild, combinations of scents probably operate at different stages in the life cycle of the fish, and they may follow a series of distinct "olfactory waypoints" during their journeys up- and downriver.[10]

As for humans, we can tell the difference between a good smell and a bad one, but few of us normally pay much attention to olfactory information, consciously at least. Vision and hearing monopolize our attention.

We can, however, make good navigational use of smells, if the circumstances are right. Approaching the coast of Luzon in the Philippines by night, I remember picking up a rich aroma of damp and decay when the yacht in which I was sailing was still far out at sea. A gentle offshore breeze was carrying it to us from the jungle-covered mountains that were still hidden in darkness. Had we been uncertain of our position, that exotic scent would have told us that we were nearing the island. Smell can also prove helpful even in much colder waters. The stink of guano can apparently reveal the presence of icebergs hidden in fog or darkness, though that is not an experience I have ever had. Advance warning of that kind must have saved the lives of quite a few sailors.

Harold Gatty, a twentieth-century master navigator, tells the tale of Enos Mills, a mountain guide, who became snow-blind while traveling alone 3,650 meters up in the Rocky Mountains, many miles from the nearest habitation. Most of us would panic if we found ourselves in such desperate straits, but Mills remained calm: "My faculties were intensely awake. The possibility of a fatal ending never occurred to me."

He could see nothing. The trails were buried beneath the thick snow, but he had in his mind a clear map of the way he needed to go. He shuffled along in his snowshoes, using his staff to find the trees, and felt the bark to locate the gashes he had blazed with his hatchet on his outward journey.

After surviving a snowslide that almost killed him, then clambering over some huge rocks and struggling through thick undergrowth, Mills caught the familiar scent of aspen smoke. As he moved steadily upwind, the smell gradually grew stronger. At last, still sightless, Mills stopped to listen for sounds of human life. It was then that he heard a little girl gently ask him: "Are you going to stay here tonight?"[11]

Darwin, sex, and hunting

Aristotle is often blamed for the contempt with which most of us treat our noses.[12] He certainly had a low opinion of our sense of smell, declaring magisterially that it was "less discriminating and in general inferior to that of many species of animals."[13] As far as he was concerned, the only use of smell was to safeguard our health by warning us when food had gone bad.[14]

But the French anthropologist and neuroanatomist Paul Broca (1824–80) should share the blame. Rather bizarrely, Broca's views about human olfaction were linked to his religious skepticism.[15] As an advocate of Darwinian ideas, Broca argued that the "enlightened intelligence" of the human being had nothing to do with the possession of a God-given soul, but rather depended on the exceptionally large size of the frontal lobes of our brains. Moreover, unlike most other animals, we were not dominated by our sense of smell, and this meant we could choose how we behaved.

So, our much-prized "free will" was merely a consequence of being not very gifted in the smell department. The Roman Catholic Church was not amused.

Broca's claim was based on the observation that the human olfactory bulb—the part of the brain that receives signals from the odor receptors in our noses—was small in proportion to the overall size of our brains. In this respect, we were very different from "lower" animals like dogs or rats that were, he believed, in thrall to their olfactory organs. It was a short but false step from there to the claim that humans had a feeble sense of smell—a view that later generations of scientists uncritically adopted. Once this piece of pseudoscience had taken hold, it was repeated again and again.

Darwin himself thought that smell was "of extremely slight service" to humans, who had, he suspected, inherited it from "some early progenitor" in an "enfeebled" and "rudimentary condition." He did, however, accept that smell was "singularly effective in recalling vividly the ideas and images of forgotten scenes and places."[16] Sigmund Freud played his own part in spreading the myth, claiming that while the sense of smell evoked instinctive sexual behavior in other animals, its weakness in humans contributed to sexual repression and mental disorders.[17]

Aristotle, Broca, Darwin, and Freud were all wrong about smell. Though some sketchy calculations made in the 1920s suggested that we could distinguish only 10,000 different odors, we can do a great deal better than that. In fact, a recent study[18] suggested that the figure should be revised upward to at least *one trillion* (10 followed by 12 zeros).

Although this finding has been challenged on methodological grounds, our sense of smell is far from being feeble. As one expert has recently observed:

> Humans with intact olfactory systems can detect virtually all volatile chemicals larger than an atom or two, to the point that it has been a matter of scientific interest to document the few odorants that some people cannot smell.[19]

According to the leading olfactory scientist, Jay Gottfried, the chemical senses—smell and taste—emerged roughly one billion years ago:

> For a bacterium tumbling through the Pre-Cambrian stew of chemicals, the sense of smell represented a keen, if rudimentary, biological adaptation, sufficing for the chemical detection of sugars, amino acids, and other small molecules. . . [and while] insects, rodents, and canines possess an unusual delicacy of smell,

even the human sense of smell astonishes: humans can tell apart two odorants differing by only one carbon atom and can detect certain odorants with an acuity better than that of rats.[20]

The human olfactory bulb may be small in relation to the overall size of our very large brains, but it is quite big in absolute terms—larger, for example, than that found in rats and mice—and it contains an unusually large number of the crucial processing units called glomeruli. In fact, though dogs have roughly ten times more smell receptors than humans, we have a larger number of glomeruli than they do.[21] Moreover, the human olfactory bulb has a hotline to the prefrontal cortex: the part of the brain that governs high-level, decision-making processes. Smell therefore differs from our other senses, all of which send their signals first to another part of the brain called the thalamus—a kind of filter that decides what deserves our conscious attention.

And that is not all. Compared to other animals, a large part of the human brain is devoted to the analysis and interpretation of the information that emerges from the olfactory bulb. We can recognize a characteristic scent even on the basis of fragmentary signals, because our brains can "fill in the gaps,"[22] and we integrate different odors into "perceptual wholes" that are laden with meaning and emotion.

Just such a process of integration was famously evoked by Marcel Proust:

> No sooner had the warm liquid, mixed with the crumbs, touched my palate than a shiver ran through me . . . An exquisite pleasure had invaded my senses, something isolated, detached, with no suggestion of its origin . . . this essence was not in me, it *was* me . . . And suddenly the memory revealed itself. The taste was that of the little piece of madeleine which on Sunday mornings at Combray . . . my aunt Léonie used to give me, dipping it first in her own cup of tea or tisane.[23]

According to the leading neuroscientist (and gastronomist) Gordon Shepherd, the extraordinarily elaborate mechanisms available to us for processing olfactory signals "bestow a richer world of smell and flavor on humans than on other animals."[24]

Lucia Jacobs is a professor of psychology at the University of California, Berkeley. She is an energetic crusader for the central importance of smell and taste, not only in humans, but right across the animal kingdom.

Jacobs told me that these two closely related chemical senses play a very important part in our lives, though we are often unaware of the influence they have on us. Women, for example, prefer male sexual partners whose immune systems differ widely from their own.[25] This unconscious bias makes good sense, as it is likely to result in healthier offspring, but it is based on differences in how men smell. What could be more important than that? And the distinct "body odor cocktails" that each of us generate also communicate information about levels of anxiety and aggression. Perhaps that explains why we unconsciously smell our hands after shaking hands with strangers.[26]

One reason why we underrate our smelling abilities is that our noses are so far from the ground. This means that we fail to notice a lot of the smells that would otherwise be available to us. But if you are willing to imitate the energetic sniffing behavior of a dog, it is quite surprising how much you can discover. Using this technique, the Botocudos tribe of Brazil and the indigenous people of the Malay Peninsula can hunt and track game, and even students in California are surprisingly adept at following a scent trail, when they get down on their hands and knees.[27]

Jacobs herself has shown that people deprived both of visual and auditory cues can identify a location by its unique odor mixture, and later find their way back to it using only their sense of smell.[28] As she comments, this is a surprising finding because "We assume that even if humans had a good sense of smell, they would not be using it for navigation, but rather for the discrimination and identification of odors."

We are "blinded by vision," as Jacobs pithily expresses it. Sight is our "default mode" and dominates our sensory world view. Our heavy reliance on it also limits our imaginative grasp of what is possible, both for us and for our animal cousins. This failure has particular relevance to the subject of navigation.

As Jacobs sees it, olfaction is the "basic command line" in vertebrates. She points out that odors are "infinitely combinatorial." This means that there is a limitless number of possible smells, each one of which could in principle act as a unique beacon or landmark. And smells that can be detected over long distances can provide an animal with invaluable directional information—perhaps even serving as the basis of a kind of map. This could be especially helpful when they find themselves in completely unfamiliar places.

∽

Lots of mammals that live on dry land seem to be good at homing—even from quite long distances. The list includes deer, foxes, wolves, polar and grizzly bears, not to mention dogs and cats.

Tracking data from seventy-seven American black bears that had been deliberately moved from their home ranges has shed some interesting light on this question. The drugged and unconscious bears were "translocated" on average a little over 60 miles—enough to take them well beyond familiar territory.

A note was made of the direction in which the bears headed on their release and they were judged to have returned home if they turned up at most twelve miles from their original capture site. The bears showed a strong tendency to head in a homeward direction, and thirty-four of them reached home before they were shot, recaptured, or their radio-collars expired. One bear made it home from a distance of 168 miles. On average, it took the bears nearly 300 days to make these journeys, but how they found their way remains frustratingly obscure.[29]

Chapter 12

CAN BIRDS SMELL
THEIR WAY HOME?

I was first introduced to the puzzling navigational abilities of homing pigeons in a university office overlooking the sunlit botanical gardens in Pisa, not very far from the famous Leaning Tower.

Paolo Luschi and Anna Gagliardo are both former pupils of the late Floriano Papi, whose work with sandhoppers we discussed earlier. Papi, who died only six months before my visit, had joined as a teenager the partisans fighting the Nazi forces then occupying Italy. He carried secret messages back and forth and would have been shot as a spy had he ever been caught. After the war, as a reward for his courageous services, Papi was given a scholarship to study at the Scuola Normale Superiore in Pisa, and went on to become an expert on flatworms, as well as light communication in fireflies. But Papi, who came from the island of Elba, was a keen sailor, and this prompted him to turn his attention to animal navigation.

Alfred Russell Wallace (1823–1913), joint discoverer with Darwin of the theory of evolution by natural selection, had proposed as long ago as 1873 that animals might find their way home with the help of smell:

> . . . the power many animals possess to find their way back over
> a road they have traveled blindfolded (shut up in a basket inside
> a coach, for example) has generally been considered to be an
> undoubted case of true instinct. But it seems to me that an animal

so circumstanced will … take note of the successive odors of the way, which will leave on its mind a series of images as distinct and prominent as those we should receive by the sense of sight. The recurrence of these odors in their proper inverse order—every house, ditch, field, and village having its own well-marked individuality—would make it an easy matter for the animal in question to follow the identical route back, however many turnings and crossroads it may have followed.[1]

Despite Wallace's prestige, other scientists did not rush to explore his idea. But in the 1970s, Papi picked up the challenge. He noticed that nobody had yet investigated the possibility that smell might play a part in the navigational repertoire of homing pigeons, though the importance of mysterious "atmospheric factors" had already been noted.

At that time, students of bird navigation were focusing their attention almost exclusively on the use of celestial cues, especially the sun compass. Birds in general were not thought to make much use of smell, or even to have particularly sensitive noses. So when Papi deprived pigeons of their sense of smell (or made them "anosmic"), and found that they were unable to find their way home from an unfamiliar location thirty-four miles west of their loft in Florence—a journey that would not normally present them with any difficulty—it came as a big surprise, even to him.[2]

Papi interpreted these puzzling results as evidence that the birds were paying very close attention to the different smells that blew across their home lofts. He thought they were associating the different scents carried to them with the direction from which the wind was blowing at the time.[3] A pigeon that recognized one of these characteristic odors at the point where it was released would set a course for home by flying in the opposite direction to the one from which the wind had carried the same odor when it was sitting in its loft. This sounds strange, but in principle it is like taking a compass bearing of a distant landmark and then, having reached it, using the reversed (or "reciprocal") course to find your way back to your starting point.

Thus was born the "olfactory navigation hypothesis." But the notion that any useful long-distance navigational information could be derived from smells was greeted with incredulity. According to Gagliardo, Papi joked wryly that even his wife refused to believe in it.

At first, almost everyone found it impossible to accept that the sense of smell could operate usefully over distances of tens of miles. One particularly serious objection was that turbulence would surely mix up the air so much as to render any long-range olfactory information hopelessly confused by the time it reached the bird's nostrils. It was also troubling that many scientists outside Italy had difficulty in replicating the results Papi had reported.

One very reasonable concern, originally shared by Papi himself, is that the procedures[4] used to deprive the birds of their sense of smell might leave them so confused or distressed that they could no longer attend to navigational cues of any kind—olfactory or otherwise. However, this seems not to be the case. Many experiments have shown that pigeons deprived of their sense of smell can navigate successfully, if they are released in a familiar area where they are able to use landmark information to find their way home.[5]

But is there any way of showing that the homing behavior of pigeons is affected by the direction of the winds reaching them in their loft?

Papi exposed young pigeons to winds that had been deflected to the left or right by vanes set up around their loft. He even tried reversing the wind direction with the help of fans. On the assumption that the winds are supplying crucial information, this piece of trickery might be expected to lead the birds astray, and that is indeed what happened. As demanded by Papi's theory, the birds exposed to deflected winds headed off in the corresponding "wrong" direction when released.[6]

There seems to be a crucial phase during their development when pigeons need access to wind information, if they are subsequently to make use of odors for navigational purposes.[7] So perhaps, rather like salmon, young pigeons "imprint" on windborne smells.

But skeptics find the "deflector loft" experiments unconvincing. Some have suggested that the vanes of the deflectors interfered with the polarized light cues on which the pigeons' sun compass may depend,[8] or that they distorted important acoustic cues.

Over the last forty-odd years, adherents of Papi's hypothesis have worked hard to address these and other objections.[9]

A leading German expert on bird navigation, Hans Wallraff, was initially as skeptical as anyone. He realized, however, that the proper response to Papi's findings was to test them thoroughly. Walraff has recently listed no fewer than seventeen different kinds of experiment that have, he believes, "yielded a coherent collection of findings supporting olfaction-based navigation."[10]

Perhaps the most striking of these involved the use of a so-called false release site.[11] Pigeons were taken in airtight containers ventilated with filtered, smell-free air to a site where they were allowed to breath the local air for a few hours, *but not released*. Then they were moved—again in purified air—to a new site located in the opposite direction with respect to their home loft. There they were deprived of their sense of smell, without having first had access to the local odors, and finally let go. The birds then headed in the "false home direction."

In other words, they went in the direction that would have made sense from the first site at which they had been allowed to sample the air but not released. By contrast, "control" birds that were exposed to the local air at the actual release site, before being rendered anosmic, followed the right course for home.

So it appeared that the first group of birds used the only information available to them—the smell to which they had been exposed at the first site—and therefore oriented in the wrong direction. On the other hand, the second group, having the advantage of the more up-to-date and relevant olfactory information, chose the right direction.

This was ingenious, but it did not satisfy everyone. Critics of Papi's theory have carried out similar "false release site" experiments, in which they exposed pigeons to nonsense, artificial odors at the "false" release site—odors that could not have provided them with any useful navigational information. They found that these birds were just as well oriented at the actual release site as controls exposed to the real, local air. They briskly concluded that "olfactory exposure provides no navigational information to pigeons whatsoever." In their view, the smells, whether nonsense or real, serve only to alert the birds to the fact that they are somewhere strange, thereby triggering some completely different navigational system; they do not supply any other navigationally useful information.[12]

More recently, however, when Gagliardo and other advocates of the olfactory navigation hypothesis tried to repeat the same experiment, they found that exposure to nonsense smells at the false release site *did* weaken the birds' homing ability.[13] It is possible that differences in the training, age and experience of the pigeons, or in their geographical surroundings, may account for these contradictory findings.

So we seem to be at an impasse, and some experts now believe that the long-running debate about olfactory navigation in pigeons will not be settled until the two sides agree to collaborate on a new set of uniformly designed experiments.[14]

Oceanic birds and navigation

Unlike pigeons, birds of the open ocean—such as albatrosses, fulmars, prions, and shearwaters—have unusually well-developed olfactory organs, which they use for finding food, for identifying their mates, and for locating their nests. Most of them are extremely long-lived (with life spans ranging from forty to sixty years) and remain faithful both to their mates and nest sites throughout their adult lives. They also cover enormous distances, and their extraordinary feats of navigation may well involve smell.[15]

Shearwaters carrying tracking devices have yielded some striking data. When temporarily deprived of their sense of smell, they have difficulty in homing, especially if released far from land. Birds taken from the mid-Atlantic island of Fayal in the Azores, and released 500 miles away, wandered for thousands of miles before they got home to their nests, while control birds that could still smell flew more or less straight back.[16]

When shearwaters were released out of sight of land in the western Mediterranean, the results were not so clear cut. The birds all managed to get home quite quickly, but while the controls followed fairly direct routes, many of those that had been deprived of their sense of smell headed north, and then followed the coastline until they reached their colony off the coast of Italy.[17] It looked as if they were searching for familiar landmarks to help them find their way home. There is also evidence that a time-compensated sun compass plays a part in the navigational toolkit of shearwaters.[18]

When shearwaters were simply tracked as they went about their normal business in the neighborhood of their colonies on the Balearic Islands, the birds that could not smell still appeared to be able to forage successfully. But the homeward routes they followed were much less direct than those of the controls—until they came within sight of the islands, at which point they could presumably pick up visual landmarks.[19] And a mathematical analysis of the paths followed by foraging shearwaters shows that they are influenced by wind speeds, in just the way that is predicted on the assumption that they rely on smell to navigate.[20]

The question then is: What smells do the birds actually use?

So far nobody has identified any naturally occurring odor on which pigeons actually rely, but oceanic birds are sensitive to certain smells that signify the presence of food, notably the compound dimethyl sulphide (DMS).[21] Of course, you cannot ask a bird what it can smell, but monitoring changes in its heart rate is a good proxy. Using this technique, Antarctic prions have been shown to be able to detect extremely low concentrations of DMS. Because this chemical plays a key role in the regulation of climate, a good deal is known about seasonal changes in its distribution, and it is known to be present in high concentrations around mid-ocean islands and above shallow underwater shelves and seamounts—places that offer abundant supplies of food.

In addition to helping oceanic birds forage successfully, the regular seasonal blooms of smelly microorganisms in places such as these may also help them find their way around. It has been suggested that entire ocean basins might present them with a relatively stable "landscape" of subtly different olfactory features, which they could become familiar with over the course of their long lives.[22]

However, the idea that a bird flying over the open sea might rely *entirely* on its sense of smell to navigate is hard to swallow, especially bearing in mind the highly turbulent nature not only of the atmosphere, but also the ocean itself.

Much of the confusion that surrounds bird navigation probably arises from the fact that birds (like many other animals) use a range of different navigational mechanisms and make their choice among them, according to the precise circumstances in which they find themselves. They may well have some way of assessing the quality of the information available from each source, before deciding which system is likely to be the most reliable, and they probably use different navigational tools at different stages in their journeys.[23]

Against this admittedly confusing background, the reader may wonder whether some completely different sense is involved in guiding pigeons (and other birds) home from unfamiliar places.

An obvious possibility is that they are making use of magnetic cues. It is well established that pigeons are sensitive to magnetic fields, but so long as they have an intact sense of smell, they generally show no sign of being disoriented when the natural magnetic field around them is disrupted by attaching magnets to their heads. And albatrosses and petrels can also home successfully under these conditions.[24] So plainly these birds do not rely *exclusively* on magnetic clues.

On the other hand, some of the procedures used to render birds anosmic apparently affect their ability to detect the presence of an artificial magnetic source.[25] So it is equally unsafe to conclude that they rely *exclusively* on olfactory cues because they cannot find their way when deprived of their sense of smell.

Are homing pigeons performing DR, or retracing their route in some other way? It is conceivable that they might use inertial mechanisms to help them find their way home, or even keep track of olfactory or acoustic landmarks, but homing performance is not much affected even when the bird is anesthetized on its way to the release site.[26] It is very hard to see how an unconscious animal can keep track of its changing course and position.

While some animal navigation experts remain skeptical about the entire olfactory navigation hypothesis,[27] many now accept that homing pigeons and oceanic birds make at least some use of smell for homing purposes. But it is still far from clear how they do so. This is a subject to which we shall return when we consider the possible role of olfactory maps.

The Atlantic puffin—with its clown-like face and whirring flight—is an irresistibly charming animal, but it is also a bit of an oddity.

While other migratory birds are loyal to a single overwintering area, puffins head off in a wide variety of directions when the summer draws to an end. And since fledgling puffins leave the nesting colony at night, apparently alone, and long before the adults, it is very unlikely that they learn the routes they follow.

Scientists who have tracked puffins from the island of Skomer, off the Welsh coast, found that in August most of them headed off first in a northwesterly direction, some going as far as Greenland, while others went south toward the Bay of Biscay. Later they all tended to move out into the North Atlantic, and then, toward the end of the winter, they moved southward, some as far as the Mediterranean, before returning to the colony in spring—from a variety of directions. What was particularly surprising was that each individual bird tended to follow the same, idiosyncratic path from one year to the next.[28]

Unlike land birds, puffins can stop at sea whenever they want, and they can probably survive the winter in a wide range of locations. So perhaps, rather than relying on any strict set of instructions—whether inherited or socially acquired—each young puffin develops its own personal migratory route, which it then faithfully follows year after year. But how they do this remains to be established.

Chapter 13

SOUND NAVIGATION

A British explorer and mountaineer named Frederick Spencer Chapman (1907–71), who was later to survive behind enemy lines in the Malayan jungle for more than eighteen months during the Second World War, was kayaking along the east coast of Greenland with an Inuit hunting party in the 1930s. There was a heavy swell running, so even when thick fog rolled in they had no difficulty in following the coast by listening to the sound of breaking waves, but Chapman could not understand how they were going to locate their home fjord. His companions by contrast were perfectly relaxed, and after an hour of steady paddling, the hunter in the lead kayak suddenly swung inshore and hit the narrow entrance exactly.

Chapman was baffled, but the explanation was wonderfully simple:

> All along this coast . . . there were snow buntings nesting, and each male bird . . . used to proclaim the ownership of his territory by singing his sweet little song from a conspicuous boulder. Now each cock snow bunting had a slightly different song, and the Eskimos had learned to recognize each individual songster, so that as soon as they picked up the notes of the bird who was nesting on the headland of their home fjord, they knew it was time to turn inshore.[1]

We do not normally navigate by birdsong, but we all rely heavily on sound to help us get around, and it can often be helpful to sailors. When approaching a high coast, a sharp sound—like a handclap or gunshot—will produce a crisp echo from a vertical rock face. Since sound travels roughly one thousand feet in three seconds, the time lag gives an indication of how close the cliffs are: a useful piece of information on dark nights or in poor visibility. And simply hearing the quality of the sound produced by breaking waves can be helpful. Waves breaking on rocks sound quite different from waves breaking on shingle, sand, or mud, and on some occasions, experienced sailors can work out where they are solely by noticing the differences.

Just like the antennae of insects, our two ears act as direction finders.[2] The minute differences in the time at which a sound reaches them, and the infinitesimal differences in its intensity, tell us whether the source is to the left or right of us. This principle underlies the 3-D sound effects created by stereo and "surround sound" loudspeakers. The change in the apparent frequency of a sound emitted by a moving source as it approaches or recedes from us—the Doppler effect—is also informative. It enables us to judge whether a car is coming toward us, for example, by listening to the noise it makes.

Blind people often make use of sound to help them get safely from one place to another. They tap with sticks or make noises with their tongues, and can tell what things are around them by detecting the subtle differences in the echoes that return to them. Interestingly, however, they often describe what they are doing quite differently. They say that they just "feel" the presence of things, and this may mean that parts of their brains not normally involved in hearing are processing the echoes.

Brian Borowski, a fifty-nine-year-old Canadian, was born blind and taught himself to echolocate around the age of three or four, by clicking his tongue or snapping his fingers:

> When I'm walking down a sidewalk and I pass trees, I can hear the tree: the vertical trunk of the tree and maybe the branches above me . . . I can hear a person in front of me and go around them.[3]

With practice it is even possible for sighted people (wearing blindfolds) to develop similar skills.[4]

Fishermen in Ghana can apparently find fish by sticking an oar in the water. The flat blade acts like a directional antenna that collects their underwater grunts and whines; by putting his ear to the handle, the fisherman can determine roughly where the fish are.[5] But the sophistication with which some animals make use of sound is truly astonishing. Bats are the best-known example.

The discovery that bats can navigate accurately in complete darkness was made in 1793 by an ingenious Italian priest, Lazzaro Spallanzani (1729–99). He had often noticed that bats came into his room at night and flew around by the light of a single candle. He decided to test their night-flying ability by catching one of them and tying a string around one of its legs. Having blown out the candle, Spallanzani released the animal and could tell by the tugs he felt on the string that it was again flying around the room, apparently quite unaffected by the complete absence of light. In further experiments (that certainly would not meet modern ethical standards), he blinded the bats and found that they could not only hunt successfully, but also find their way back to the bell tower where he had captured them.[6]

Spallanzani's discoveries went largely unnoticed at the time, as very few of his findings were published. It was not until 1938 that the nature of the bat's night-flying abilities was explained by a young Harvard researcher named Donald Griffin (1915–2003), who was interested in their seasonal migrations. He and his colleague Robert Galambos were able to show that bats can detect flying insects and home in on them in the dark by emitting ultrasonic clicks and buzzes, and analyzing the returning echoes: a system very like the sonar used to hunt submarines. Griffin realized that their extraordinary navigational and prey-catching abilities must depend on the construction of a highly detailed, three-dimensional "view" of their surroundings.[7]

Moths are an important part of the diet of bats, and some of them have developed their own countermeasures. They perform evasive maneuvers when they pick up the special signal used by a bat as it closes in on its prey, or even "jam" the bat's sonar by emitting a signal of their own, so bats have to be extremely nimble to catch them.[8]

Echolocating bats have a good claim to be the master navigators of the mammalian world, and the challenges they face are enormous. In the first place, they need to be able to work out where they are and what is around them solely by listening to the echoes from the sounds they are emitting. Imagine what this means: They have to identify a barrage of different sounds reflected from every surface around them—a grassy meadow, the bark of a tree or its leaves, a brick wall, a tiny flying insect, or the surface of a pond.

That would be hard enough if the bat were stationary, but bats can fly very fast and seldom go in a straight line; in fact, their midair maneuvers are more impressive than those of most birds. And to make matters even more complicated, they may also have to distinguish their own signals from those of other bats of the same species flying around them.

Flying in complete darkness, some bats can reliably find a tiny hole in a narrow grid of wires and fly safely through it. Others follow regular "flyways" from their roosts to their hunting grounds every night, which take them through tortuous underground passageways that extend over several miles.[9] But echolocation has its limitations: Its maximum effective range is only about 100 meters, so it is of no help in detecting distant landmarks. For long-distance navigation, bats therefore have to fall back on other senses—especially vision (*see page 27*).

Sonar is used by other mammals—notably dolphins, porpoises, and other "toothed" whales—for tracking down and capturing their prey.

Captive dolphins are extremely adept at detecting small targets underwater, even in complete darkness, and can certainly use sound to avoid obstacles. The high-intensity, ultrasonic clicks they produce give them a picture of their environment stretching out to around 300 meters, and radio-tracking studies conducted in the open sea suggest that they use this system to follow the underwater topography.[10] A study of two captive porpoises also showed that they employed their sonar to orient themselves in relation to landmarks.[11]

There is not much hard evidence that whales and dolphins use sonar for navigational purposes, but it would be surprising if they did not. In fact, some researchers believe that the original purpose of their sonar system, like that of bats, may have been navigational.

It is tempting to speculate that whales on their long-distance migratory journeys may make use of the underwater "auditory land-scape." Though the signals they emit are probably not strong enough to provide much useful information when they are traveling in the deep ocean (which is typically around two miles deep), they could be helpful in shallower seas and around seamounts.

The Concorde effect

Jon Hagstrum is a geophysicist at the US Geological Survey, and he has been struggling for the last twenty years or so to persuade the world that the pigeon has a sophisticated navigational system that relies on low frequency sound, otherwise known as infrasound. The fact that he is not a professional biologist may at first seem puzzling, but his unusual professional background qualifies him rather well to explore this particular question. I interviewed him at his office in the suburb of Menlo Park, near Stanford University, just south of San Francisco.

Hagstrum's father was a physicist and wanted him to follow in his footsteps, but his son was determined to follow a career that would give him a challenging outdoor life. Being a photographer for the *National Geographic* magazine would have been ideal, but he decided to go down the slightly more realistic path of studying biology at Cornell University. The course there was designed for medical students, and when he discovered how much time he was going to be spending in a laboratory, he switched to geology. In 1976, he happened to hear a talk by Bill Keeton (1933–80), one of the leading researchers in pigeon navigation at that time.

Hagstrum was gripped by what Keeton had to say, especially about the odd behavior of certain pigeons released in the neighborhood of a place called Jersey Hill. These birds were invariably disoriented and very rarely managed to home successfully. And they had one thing in common: They all came from a loft at Cornell. Strangely, birds from other lofts in upstate New York released at the same site were unaffected. Keeton had struggled to come up with a plausible explanation for this bizarre phenomenon and, turning to the audience, asked if anyone had any bright ideas. This rhetorical challenge caught Hagstrum's imagination and he never forgot it.

A few years later, an article in *National Geographic* reignited Hagstrum's interest in the problem, and he was struck by how little had been done to see whether sound might be the missing clue. He had by now taken classes in seismology, so he knew a lot about how sound waves propagate, and he had read more about animal navigation, too. But his career as a geophysicist kept him busy traveling all over the US, and he was unable to take the matter any further. Finally, in 1998, Hagstrum read some articles about pigeon races in the eastern US, as well as in Europe, that had been mysteriously "smashed": the technical term used when the birds fail to reach home in good time, or even at all.

It was well established that pigeons had access to two kinds of compass—one solar and one magnetic—but a compass by itself will not enable a bird to home successfully from somewhere it does not already know. It also needs a map of some kind. One widely discussed theory is that birds may use gradients in the intensity of the earth's magnetic field as the basis for such a map. Hagstrum was convinced that this solution was unworkable, but he was also deeply skeptical about Papi's olfactory map hypothesis. And in any case, neither theory could satisfactorily account for what Keeton had observed again and again over a period of nearly twenty years at Jersey Hill.

Hagstrum found himself irresistibly drawn to the notion that sound might be the key—a possibility that Griffin (of bat echolocation fame) had also speculated about many years earlier. Paraphrasing a famous remark attributed to the great physicist Neils Bohr, Hagstrum thought "maybe that idea's just crazy enough to be right."

The sounds we can hear do not carry very far in air, but some animals are sensitive to very low frequency sounds that are well below the threshold of human hearing (~20 Hz). This so-called infrasound is dissipated far more slowly and can travel for thousands of miles. In principle, it ought to be possible to orient by reference to such signals.

Homing pigeons can certainly detect infrasound,[12] though why this ability should have evolved in the first place is uncertain. One possibility is that pigeons, and perhaps other birds, make use of infrasound to detect the approach of weather fronts that will bring strong winds and rain. That would be a valuable asset for any bird that undertakes long journeys.

Could Hagstrum find some sort of acoustic disturbance—probably infrasonic—that might have confused the "map sense" of the birds in the "smashed" races?

He unsuccessfully explored a number of different possibilities before he found what might be the answer: the boom produced by the Concorde supersonic airliner (which was still in operation at the time). Could this very powerful source of infrasound have overwhelmed the pigeons' navigational system, or perhaps temporarily deafened them?

Hagstrum discovered that more than 60,000 pigeons from home lofts in England had been "tossed" (the unceremonious launch of the pigeon into the air) at Nantes in northern France on June 29, 1997, in a race marking the centenary of the august Royal Pigeon Racing Association. Normally, 95 percent of them would have been expected to return safely, but on this occasion very few made it. It was such a disaster that an inquiry was launched, but the eventual report was inconclusive. The puzzled race organizers attributed the losses to the usual suspect: bad weather.

But Hagstrum calculated that most of the pigeons would have been over the English Channel at exactly the time the daily Concorde flight from Paris to New York was passing above them, having gone supersonic after crossing the French coast.[13] And significantly, the few birds that did home successfully were slow fliers, which meant that they had not reached the sea by that time. So, this looked like a promising explanation.

Next, Hagstrum looked at data from several disrupted races that took place in 1998, one in France and two more in the US. Although it turned out that the birds taking part in these races could not have encountered the cone-shaped shock wave that surrounds the aircraft when flying supersonically, the timings (and weather conditions) meant that they could have met the slower-moving acoustic waves as they moved on ahead of the aircraft when it decelerated before landing.

There was, however, one exception: a smashed race in Pennsylvania. When Hagstrum examined this event, he discovered that the scheduled arrival time of Concorde would have been far too early. There was only one possibility remaining and it was a long shot. If his theory was correct, Concorde had to have arrived in New York more than two hours late that day. So Hagstrum called Air France at JFK Airport to find out. The official to whom he spoke was at first sniffily dismissive of the idea: How could the mighty Concorde possibly have arrived so late? But when Hagstrum explained that his inquiry was scientific in nature, he reluctantly agreed to check.

When Hagstrum called back later, the Air France man asked, "Are you a magician?" Mechanical problems in Paris had indeed delayed the aircraft that day by two and a half hours, so the Pennsylvania birds could have encountered the shock waves after all. Hagstrum points out that the behavior of the pigeons had enabled him not only to predict the delay of the aircraft, but even its duration. It was, he says, perhaps the most exhilarating moment of his scientific career, but he still had difficulty in getting his findings published.[14]

Hagstrum has more to go on than the fact that a few smashed races coincided with Concorde flying by. He has also studied records of 2,500 releases involving 45,000 birds dating from Keeton's time at Cornell. Keeton was a highly respected scientist, and the fact that his data are not new in no way weakens their significance. In fact, it removes the possibility that Hagstrum himself might have introduced some unconscious bias.

As we have seen, Keeton found that birds from home lofts at Cornell, when released at Jersey Hill, typically headed off in random directions and that only about 10 percent actually made it home. At Castor Hill, the story was rather different but equally odd. The birds released there usually went off in a consistent direction, but it was often the wrong one. At another release site, near Weedsport, the birds almost invariably homed accurately, but on one exceptional occasion they failed to do so. Could one process explain all of these bizarre results?

Infrasound is generated by a variety of natural processes, including storms at sea and tornadoes on land, as well as interactions between high winds and landscape features like mountains. Waves breaking on a shore are also a source. Standing waves in the open ocean are, however, particularly significant. Sometimes also called stationary waves, these are like the stable wave patterns that appear on the surface of a cup of coffee when you bang repeatedly on the table beneath it; similar waveforms also underlie the tones produced by musical instruments.

The standing waves that interest Hagstrum are on a far grander scale. They are caused by constructive interference between the colossal wind-driven waves produced by storms or hurricanes on the open ocean, and occur when two wave trains of similar frequency, but traveling in opposite directions, encounter one another. Such standing waves give rise to oscillating changes in air pressure (called *microbaroms*), which can propagate upward all the way to the stratosphere.

Up there, temperature gradients and fast-moving currents of air can bend them toward the surface, where they are reflected upward again. As this process is repeated, it creates a "waveguide"—a sort of sonic pipeline—that can channel the microbaroms over huge distances.

And that is not all. The very same standing waves also generate tiny earthquake-like vibrations (called *microseisms*) in the ocean floor beneath them. These radiate outward until eventually they can be detected by seismometers right in the heart of continental landmasses.

In fact, microseisms and microbaroms from this oceanic source produce an almost continuous background hum of infrasound in the solid earth and atmosphere, respectively, each with a frequency of ~0.2 Hz and a period of around six seconds. They cause real problems for scientists trying to detect other important things, like distant earthquakes, or the signals generated by nuclear tests.

Hagstrum has proposed that pigeons can detect the infrasound that arises from slight oscillations of the land surface, and that this ability underlies their extraordinary homing ability. To be more precise, he thinks that each pigeon learns to associate its home loft with a kind of infrasonic signature, or sound print, shaped by the landscape features that surround it. He is not sure whether microbaroms traveling in the air, or microseisms passing through the earth and reaching up into the atmosphere, dominate the underlying process (although most evidence points to the latter: see below).

In any event, the signature sound radiates out from the home loft's neighborhood, rather like the note of a bell (only much, much lower and completely inaudible to us). Sounds at these ultra-low frequencies can travel over very long distances in the air and may act like a beacon, enabling the pigeon to set an accurate course for home—in normal circumstances.

But if the sound is diverted by atmospheric temperature gradients, or landscape features, the pigeon is going to be in trouble, and this is how Hagstrum interprets the peculiar behavior of the birds at Jersey Hill, Castor Hill, and Weedsport.

Ulysses S. Grant and zones of silence

With the help of a computerized, atmospheric modeling program, Hagstrum has shown how the propagation of infrasound is affected both by the temperature and wind structure of the atmosphere, and by changes in the weather, as well as the physical shape of the landscape. Such factors can give rise to local "zones of silence," or acoustic shadow zones, within which pigeons would be unable to pick up the crucial sound signatures emanating from the neighborhood of their home loft.

Zones of silence caused serious problems during the American Civil War. Army commanders on both sides often kept large forces in reserve, and only committed them when sounds of battle indicated that they were needed. Sometimes, however, despite being near at hand—they heard nothing. Acoustic shadowing probably explains why General Ulysses Grant failed to reinforce his subordinate, General Rosencrans, at the Battle of Iuka on September 19, 1862. The thunder of the guns simply had not reached him.[15]

Atmospheric modeling shows how an acoustic shadow zone could have given rise to the strange disorientation of the Cornell birds released at Jersey Hill. In normal circumstances, infrasound from Cornell does not reach Jersey Hill. Yet there was a single occasion when Cornell birds were able to home successfully from this location. Hagstrum has shown that unusual weather conditions on that particular day would have radically altered the way in which infrasound from Cornell propagated. As a result, birds at Jersey Hill would have enjoyed a rare day of acoustic contact with Cornell and would have been able—for once—to pick up the beacon of their home loft.[16]

The misorientation of the Castor Hill and Weedsport birds, on the other hand, may have been the result of infrasound signals arriving from more than one direction, in response to varying weather conditions and terrain features that favored different directions of propagation. Some other rare anomalies that emerge from Keeton's data can even be explained by the interfering effects of infrasound from tornadoes and hurricanes, which were recorded on the relevant days.

One oft-repeated objection to Hagstrum's hypothesis is that the pigeon's ears are so close together that the bird could not possibly extract any useful directional information from low-frequency sounds with wavelengths of a half mile or more. If the bird were unable to move, this would be a telling criticism, but by flying in a circle or a loop, it can artificially extend the size of its listening apparatus. Taking advantage of the Doppler effect, it can then determine the direction from which the sound signature of home is reaching it.[17] Radar engineers employ exactly the same principle and call it a "synthetic aperture." The fact that pigeons at the point of release often fly in circles or loops before heading off in a homeward direction is consistent with the notion that they are extracting directional information from infrasound.

A much more serious objection is that surgically deafened pigeons are still able to orient toward their home. But the evidence here is neither strong nor clear. The first study of this kind was small in scale and the results were not consistent: Some deafened birds failed to orient, while curiously some intact control birds were also unable to do so.[18]

Hagstrum has recently reviewed an unpublished set of data, generated again by Keeton, that sheds further light on this subject. The deafened birds in Keeton's various tests—taken as a group—did behave differently from the controls, and were generally less well oriented, though again many of them did manage to home successfully. But some control birds, with intact hearing, were also unable to orient.[19]

Hagstrum believes that the controls were at times victims of acoustic shadows. The deafened birds (realizing that their hearing was disabled) may have been monitoring one of their compass senses on their outward journey from the home loft, as young inexperienced pigeons do, and then flew home on a reverse course. Another piece of indirect evidence comes from a curious seasonal pattern in the homing abilities of European pigeons. In the winter, they tend to be less well oriented and slower in homing than in summer. This anomaly is known in German as the "Wintereffekt," but it has not been observed in North America. Hagstrum suggests that it is due to an increase in background infrasound noise caused by the larger number of storms over the North Atlantic in winter, which is preferentially ducted toward Europe (rather than America) by the stratosphere's westerly winds.[20]

Proponents of the olfactory navigation hypothesis point out that the Wintereffekt might also be explained by the reduction in the availability of navigationally useful smells produced by plants in the winter months.

Hagstrum would be the first to agree that anecdotes like these provide only circumstantial evidence in favor of his infrasound hypothesis. His own commitments make it hard for him to carry out the kind of experiments that are necessary to establish definitely whether or not pigeons make use of infrasound, but he hopes that others may soon be able to do so.

∾

A lot of animals return to their birthplace, when the time comes for them to breed, but it is difficult to study the detailed behavior of animals—like seals— that breed in large colonies, not least because they will attack if you go too close to them.

Recently, scientists have overcome these difficulties at a major Antarctic fur seal colony on Bird Island, off the coast of South Georgia.[21] Here an elevated walkway makes it possible to locate individual seals with great precision. With the help of electronic identification tags that could be read with a device on the end of a long pole, the researchers found that female Antarctic fur seals return with extraordinary precision to the places where they were born, to give birth to their own pups—even after an interval of several years.

Most came back to within twelve meters of their birthplace and some even returned to within one body length (two meters). Though the males— whose numerous mates form "harems"—have not yet been studied in the same way, it is possible that they show even greater site fidelity. Photos of an Alaskan fur seal colony taken in the 1890s show "virtually the same pattern of distribution of harems as today."

Nobody knows how the seals home with such precision. Perhaps sight and smell are important in the very last stage of their homing journey, after they come ashore, and maybe they use celestial or magnetic cues when they are far out at sea. Who knows?

Chapter 14

THE EARTH'S MAGNETISM

For hundreds of years, mariners have relied on the magnetic compass to enable them to set a course and steer by it. Learning to recite the thirty-two "points" of the compass was a rite of passage for every sailor, until they were replaced with degrees. Now north is simply zero, east is 90, south is 180, and west 270—and so on up to 359. Though you might still say "southeast" or even "north-northwest," the more complicated "points" are largely forgotten.

Lodestones—pieces of permanently magnetized rock (magnetite) that draw iron toward them—were described in antiquity, and their tendency, when freely suspended, to "seek the north" must soon have been discovered. The Chinese seem to have invented a kind of compass about two thousand years ago. It is not clear when they began to make use of this wonderful new tool for navigational purposes, but they had certainly begun to do so by the eleventh century. By the twelfth century, the compass had also made its appearance in Europe, though whether it was separately invented in Europe is still a subject of debate.[1]

The European voyages of discovery that began in the fifteenth century depended as much on the compass as on the instruments used to measure the height of the sun and stars, and the transoceanic trade routes that soon developed could not have been sustained without its help. A case can be made that it was the single most important navigational tool until the coming of GPS, and even today, no ship is complete without its steering compass.

Deep beneath our feet, surging vortices of molten metal heated by the solid heart of the earth (which has a temperature of almost 6,000 degrees Celsius, or about 11,000 degrees Fahrenheit) generate a magnetic force field that envelops the entire planet.[2]

Without the protection of this geomagnetic field, as it is called, there would be no life on our planet. Reaching far out into space, it deflects the high energy particles streaming out from the sun that would otherwise rip apart the atmosphere's protective ozone layer. Everything would then be bathed in a deadly solar flux that would sterilize the earth's surface.

The geomagnetic field is like that surrounding an ordinary bar magnet, though obviously on a much grander scale. It has two poles joined by looping "lines of force." The magnet inside a compass aligns itself with these: One end points to the north, while the other points to the south geomagnetic pole. In other words, it is sensitive to *magnetic polarity*. But, as mentioned earlier, there is a problem: The magnetic poles seldom coincide with their geographical counterparts. In fact, they are currently hundreds of miles away from them, and they are constantly on the move.[3]

As a result, almost everywhere on the surface of the earth true (geographical) north/south differs from geomagnetic north/south. The angular difference between the two goes by the technical name of *declination*.[4] It is vital to take account of declination when navigating with a compass, or you may end up somewhere quite unexpected. And if you are anywhere near either of the geomagnetic poles, where the declination changes very rapidly over quite short distances, a magnetic compass is practically useless.[5]

The brilliant English astronomer Edmund Halley (1656–1742), of comet fame, set sail in 1699 on a long and difficult voyage in the Atlantic and Indian Oceans (even reaching the edge of the Antarctic ice pack), during which he made a series of measurements of magnetic declination. On his return, he published an elaborate chart on which lines of equal declination were recorded, in the hope that it would prove helpful to mariners in determining their longitude. It was a good

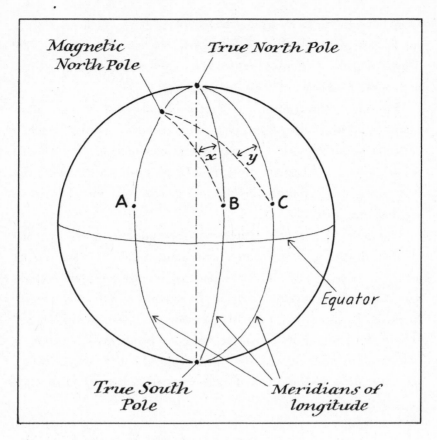

Magnetic declination varies greatly from place to place. At point A declination is zero. Angles "x" and "y" show the declination values at points B and C.

idea in theory, but it failed to catch on. While Halley had proved that it was possible to determine magnetic declination at sea, it was not at all easy to do so accurately, and there was the further problem that declination values were constantly changing. As a result, Halley's chart—though a remarkable cartographic achievement—was never widely used.

The lines of force that emerge vertically from one pole and descend again vertically to the other flatten out as they sweep around the earth and run parallel with its surface in equatorial regions. The variable angle between the geomagnetic field and the earth's surface is known as its *inclination*. Mariners use the more descriptive term *dip*, and it is easy to see why. If a magnetic needle is allowed to pivot in a vertical plane, it will remain level when close to the equator, but one end will gradually dip more and more steeply below the other, as it nears the pole that attracts it.

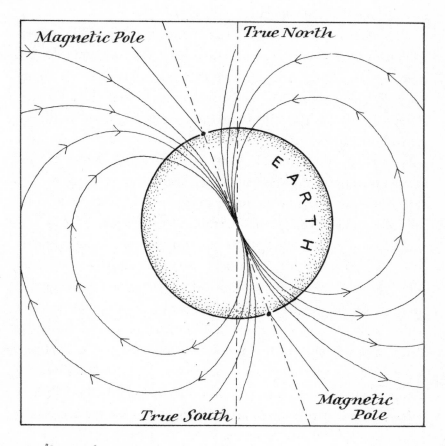

The "lines of force" that link the north and south magnetic poles

Magnetic inclination is navigationally helpful but ambiguous. It steadily increases as you approach either magnetic pole, while it decreases as you approach the equator, but it cannot tell you whether that pole is the north or south one.

Another important feature of the geomagnetic field is its strength or *intensity*. This is greatest near the poles and gradually weakens as you approach the equator, though it varies much less (and more irregularly) in an east-west direction. In absolute terms, it is not at all strong. Measured in nanoteslas (nT), it varies between about 25,000 and 65,000—for comparison, a small fridge magnet exerts a force of about 10,000,000 nT.

The intensity of the earth's magnetic field is also highly uneven and varies over time. Every year, every day, every hour, indeed every minute, the field intensity changes from place to place in a highly unpredictable fashion. The longer-term fluctuations ("secular variation") arise from the still partly mysterious processes unfolding around the earth's core, while the rapid changes occurring over the course of a day result from electrical activity in the ionosphere caused by its exposure to the sun. Allowance also has to be made for the three-dimensional nature of the geomagnetic field, since its intensity falls off rapidly as you rise above the earth's surface.

The complex and highly dynamic nature of the earth's magnetic field means that any two-dimensional map of intensity gradients can offer only a very rough approximation of the actual values in any given place.[6] Magnetized rocks in the earth's crust, the presence of which is erratic, also give rise to local fluctuations in field intensity that can overwhelm the background north-south gradients. These local anomalies are sometimes powerful enough to disrupt a steering compass and are therefore marked on marine charts. For all these reasons, it is very difficult to obtain reliable information about your location simply by measuring magnetic intensity.[7]

There is one more factor to be considered. Every now and then, at irregular intervals, the entire geomagnetic field upends itself: The north pole becomes the south pole and the south, the north. The last time a major polarity shift occurred was about 780,000 years ago, but many more such "field reversals" have occurred in the more distant past. We know about these events from the fossilized traces they have left in rocks on the sea floor.

Typical reversals are thought to take several thousand years to complete, and during this interval, the decay of the preexisting twin-pole field may give rise to bizarre multi-polar fields. This would be quite disconcerting if you were relying on any kind of magnetic compass to set your course.

The inclination compass

The idea that animals might make use of the geomagnetic field to navigate was much discussed during the nineteenth century. The Russian zoologist and explorer Alexander von Middendorf (1815–94) raised the possibility in 1855; and then in 1882, an obscure Frenchman living in Algeria called Viguier examined the ways in which animals might be able to exploit both magnetic inclination and intensity to find their way.[8] He presciently described a possible experiment in which pigeons could be fitted with either magnetic or nonmagnetic bars, in order to see how these might affect their homing ability.

But the idea did not take root, and the magnetic navigation hypothesis was largely ignored by the scientific establishment until the 1960s. By then, a steady trickle of discoveries had encouraged hitherto skeptical researchers to think again. Evidence was emerging that a surprisingly wide range of animals—including termites, flies, sharks, and snails—were sensitive to magnetism, and soon the list was extended to include honeybees and birds.

The first indication that bees could detect magnetism came from an experiment in which the natural field around the hive was cancelled by a magnetic coil system. The direction indicated by the foragers in their waggle dance then changed very slightly. More intriguing still was the discovery that the seemingly disoriented dances performed by bees deprived of any celestial cues (either the sun or e-vectors) actually followed a pattern: They generally preferred to indicate directions that matched the four cardinal points of the magnetic compass. When the magnetic field around them was canceled, this curious "nonsense" pattern disappeared.

The honeybee plainly can detect the earth's magnetic field, but it may not make direct use of it for navigational purposes. More probably it uses the regular daily changes in the earth's magnetic intensity that occur around sunrise and sunset to calibrate the internal clock that governs its sun compass. And perhaps other animals do so, too. The bee's magnetic sense also helps it to construct a hive with an orderly array of cells. What remains uncertain is whether it uses magnetic information to help it navigate when the sky is obscured by cloud and its sun compass is out of commission.[9]

Signs of a magnetic sense in birds began to emerge in the 1960s, as a result of pioneering work by Friedrich Merkel and Wolfgang Wiltschko.[10] But the great breakthrough came from a crucial experiment conducted by Wolfgang Wiltschko and his wife, Roswitha, in 1971.[11] They placed migratory European robins in an octagonal cage, with eight perches equally spaced around its perimeter. They then exposed the birds, which were in the grip of *Zugunruhe* (the delightful German word for the restlessness that birds display when they are about to migrate), to altered magnetic fields, and recorded which perch they chose to settle on. Their aim was to see which components of the field shaped the birds' behavior: its intensity, inclination, or polarity.

The Wiltschkos systematically reversed these parameters in a series of different combinations. What they found was quite surprising. The birds' directional preferences depended not on the *polarity* of the field, but rather on its *inclination*. They could therefore judge which direction pointed toward the nearest magnetic pole, but could not distinguish between north and south. So their compass was quite unlike the kind we humans are familiar with. But this does not mean that they can only fly north or south; once their magnetic compass is calibrated, they can set a course in any direction they choose.

An "inclination compass" of this kind would be expected to work well in mid to high latitudes, where the angle of dip, being reasonably steep, should be readily apparent to the bird. But it will be ambiguous when the field lines are horizontal, as is the case near the equator, and that is exactly what the Wiltschkos found. When tested in a horizontal field, the robins had no idea where to go and became disoriented. This finding has important implications. It means that birds migrating from the northern to the southern hemispheres (and back) cannot rely on their compass sense when they approach the magnetic equator.

The Wiltschkos' findings have since been replicated in many different laboratories and count as one of the most significant discoveries in the history of animal navigation studies.

An inclination compass has been found in twenty different species of birds (as well as many other animals), and it may be a universal avian gift. In some migratory birds, it seems to be the prime orientation mechanism in daylight, though it is calibrated by access to skylight polarization patterns.[12] Nocturnal migrants can also use an inclination compass, which they calibrate by the azimuth of the sun at twilight—a technique that enables them to maintain a steady heading, even when crossing the equator.[13] But the level of precision offered by the inclination compass is the subject of some debate. A long-distance migrant certainly could not rely on this mechanism alone to reach a small target area, like an island in mid-ocean, as a compass can give no warning of sideways drift.

As more and more studies are published, it is becoming increasingly clear that the magnetic-compass sense is not a rare phenomenon.[14] In addition to birds and reef fish,[15] a variety of invertebrates including fruit flies[16] and beetles[17]—to name but a few examples—also seem to have magnetic compasses.

Humpback whales make long journeys from their summer feeding grounds in the chilly but food-rich seas around Antarctica to the warm tropical waters of the central Pacific and Atlantic Oceans, where the females give birth to their calves—a distance that may exceed 5,000 miles.[18]

More remarkable still is the precision of their navigation. In a recent tracking study,[19] humpbacks in both the Pacific and South Atlantic followed arrow-straight tracks across the ocean, often for days on end. They were clearly able to compensate for the effects of crosscurrents, and in one case the passage of a tropical storm—a disturbing event even for a large whale. This is no mean feat, though nobody knows what cues they rely on and it is very difficult to conduct experiments on them, apart from tracking ones, both on practical and ethical grounds.

It is possible that whales make use of magnetic cues, and the observation that they beach themselves—sometimes in the hundreds—is taken by some researchers as evidence that they are sensitive to magnetic fields. These "mass strandings" are often fatal for the individuals concerned, and they have long puzzled scientists.

Many possible explanations have been proposed, including the disturbing effects of loud underwater noises caused by human activities.[20] But strandings on the east coast of the US seem to be focused on areas where the magnetic intensity is relatively low, so it is possible that intensity gradients play some part in their guidance system.[21] Following a similar line of reasoning, other scientists suspect that recent sperm whale strandings in the southern North Sea were caused by a powerful solar storm that disrupted the earth's magnetic field.[22]

There are, however, many other possible explanations. Perhaps whales make use of the sun, moon, or even the stars to maintain a steady course—they quite often raise their heads out of the water as if to look around them (a behavior known as "spy-hopping"). There is also evidence that they like to visit underwater features called seamounts, and these may perhaps act as navigational beacons for them.[23] Passive listening, echolocation, olfaction, and perhaps even gravitational gradients may also be involved.

Chapter 15

SO HOW DOES THE
MONARCH NAVIGATE?

L et me now return to my childhood inspiration: the annual migration of the North American monarch. The true nature of this
extraordinary phenomenon remained a mystery until surprisingly
recently. One man, a determined Canadian entomologist called
Frederick Urquhart (1911–2002), deserves much of the credit for
solving it.[1]

From his childhood days, Urquhart was mad about moths and
butterflies, and the monarch inevitably caught his attention. The fact
that these insects disappeared during the winter months was well
known, and there was evidence that some of them headed south, but
it was unclear how far they went, and there was also the possibility
that some of them might be hibernating (presumably in well-hidden,
sheltered locations).

Though Urquhart searched diligently in the hope of finding a dormant monarch, he never succeeded in doing so. Where then were they
all going? This question still haunted him when he was a graduate
student at the University of Toronto during the Depression years of
the 1930s, but he was only able to continue his investigations in his
spare time—with his mother's "able and enthusiastic assistance."[2]

During the Second World War, Urquhart was posted as a meteo-
rologist to various parts of Canada and was able to make surveys of
the local monarch populations, but it was only in 1950 that he was able

to obtain funding for a serious program of research. Now helped by his wife, Norah, he embarked on the project that turned out to be his life's work. Since it is so difficult to track butterflies visually—and impossible to do so over long distances—the Urquharts decided to try tagging them.

This did not prove easy to do, but they devised a way of gluing small paper tags to monarchs, each bearing a unique number and a request for the finder to send in a report. The process involves gently holding the insect and scraping off a small patch of the microscopic scales that cover its wings so that the sticky tag stays attached. Apparently, this does not much trouble the butterflies, though it must call for a good deal of dexterity on the part of the tagger.

In 1951, Norah wrote an article about tagging monarchs that excited the interest of many naturalists and biologists. As a result, Urquhart was "deluged with offers of assistance from all over the United States and Canada."[3] More than 300 volunteers signed up as "cooperators." It was an early and highly successful example of citizen-science crowdsourcing.

With the help of this small army of volunteers, the Urquharts laboriously caught and tagged more than 300,000 monarchs. Reports of sightings began to trickle in and a pattern emerged. Most butterflies that were tagged east of the Rocky Mountains (there is a separate population on the Pacific side that behaves differently) seemed to head south into Texas, and then over the border into Mexico. The Urquharts were eventually able to trace the migrants as far as the volcanic mountain ranges west of Mexico City, but there the trail went cold.

It was not until the 1970s that the almost monomaniacal perseverance of the Urquharts finally paid off. Unable to make any further progress using tags, they put advertisements in the Mexican papers in the hope that someone might help them fill in the last piece of the puzzle.[4]

In 1973, an American living in Mexico City, Ken Brugger, saw one of their ads and went hunting for monarchs in his motor home with his Mexican partner, Catalina Aguado. Two years later, high in the mountains, the couple was caught in a hailstorm, but hail was not the only thing falling from the sky—there were also thousands of battered monarchs. They soon discovered the first of the over-wintering sites that had for so long eluded the Urquharts. Here literally millions of butterflies clustered so thickly on the firs, pines, and cedars that the trees were bent over under their weight, and the floor of the forest was thickly carpeted with dead butterflies that made a feast for the local cattle.

The Urquharts themselves visited the site as soon as they could and even managed to find a few butterflies with tags attached to them. This was the crucial proof they were looking for: At least some of the insects clinging to the trees had indeed traveled south from the US. Later research, relying on measurements of carbon and hydrogen isotopes in the butterflies' wings, has enabled scientists to locate the feeding grounds of the caterpillars from which they are descended. Most of the monarchs in their Mexican mountain retreats come from the American Midwest.

When he announced the amazing discovery in 1976, Urquhart was tight-lipped about the precise location of the overwintering site. He revealed only that it lay on "the slope of a volcanic mountain situated in the northern part of the State of Michoacan, Mexico, at a height of slightly over 3000 meters." Urquhart was no doubt afraid that too much public attention might threaten the vulnerable butterflies, but he refused even to share the details with a fellow lepidopterist, Lincoln Brower, another devotee of the monarch. In fact, he went so far as to lay a false trail for him.

But Brower was not fooled. From clues inadvertently left by his unforthcoming colleague, he managed to work out where the original site was, and by 1986 he had found eleven more. The first alone contained upward of 14 million butterflies in an area of 3.7 acres. All the sites stand, or stood, in forests at altitudes of around 3,000 meters, where the butterflies enjoy cool but stable conditions in which they pass the winter months in a quiescent state known as diapause.

Urquhart's extraordinary revelation, which made front-page news around the world, came as no surprise to the locals, who had long known about these extraordinary gatherings of butterflies. And today, though much diminished in scale and number, the overwintering sites are a popular tourist attraction.

When the days lengthen in the spring, the butterflies get sexually excited and rise up in vast numbers from the trees. The males sprinkle an aphrodisiac dust on the females and grapple them to the ground. After their mating frenzy, the butterflies stream northward, many of the males dying off along the way. The females lay their eggs on milkweed plants in the southern US and die. The caterpillars hatch, feed, and eventually pupate.

Then a new generation of adults emerges and heads farther north, where the females once again lay their eggs. At the end of the summer, in response to the shortening days, the last brood of adult butterflies (the fourth or even fifth) heads south to Mexico. Some of them are in Canada when they start out on their long trek. They may fly as much as 2,200 miles over a period of seventy-five days—roughly thirty miles a day. But these insects have never made the journey before and there is no one to show them the way.

When heading north, the female monarchs have a relatively straightforward navigational challenge. They need only find the milkweed plants and lay their eggs. But when the shorter, colder days of autumn announce that it is time to head south, both males and females need to find their way back to the distant, isolated overwintering sites.[5] Although it is hard to imagine how such a feat is possible, a series of remarkable discoveries over the last twenty years or so have transformed our understanding of it.

Antennal circadian clocks

Drawing her inspiration from the earlier work of von Frisch and Wehner, Sandra Perez (of the University of Arizona) decided in the 1990s to find out whether the monarch, like the honeybee and desert ant, made use of a sun compass. Employing the so-called clock-shift technique, she kept one group of monarchs in a room, where the lights were turned on and off to simulate a day starting and finishing six hours later than the natural day. One control group was kept indoors, but not subjected to any clock-shift, while another consisted of captured wild butterflies that had not been cooped up at all.

Perez and her energetic colleagues released the butterflies one by one and estimated their heading with the help of hand-bearing compasses while running alongside them.[6] When they compared the average headings of the different groups, they found that the clock-shifted butterflies headed in a west-northwesterly direction, while the controls all followed the normal south-southwesterly courses.

This was exactly what was to be expected if the butterflies were using a time-compensated sun compass. Perez also noted that the monarchs seemed to be able to maintain their heading when the weather was overcast. She therefore thought they might have a "non-celestial" backup compass, perhaps based on the earth's magnetic field.

Some years later, Henrik Mouritsen, a leading animal navigation scientist based at the University of Oldenburg in Germany, and his colleague, Barrie Frost, of Queen's University, Kingston, Ontario, found a way of monitoring the orientation of insects in flight with more precision—and much less running about.[7] It involved tethering the butterflies in a kind of flight simulator, which allowed them to monitor and record their heading for as much as four hours at a time (the equivalent of flying a distance of roughly forty miles).[8]

Mouritsen and Frost clock-shifted two groups of butterflies: One was six hours "fast" and the other six hours "slow." The control butterflies reliably oriented, more or less as Perez had noted, in a southwesterly direction. In fact, their average course matched remarkably well the route that would eventually have taken them to their destination in Mexico.

The orientation of the two clock-shifted groups was also highly consistent: the "fast" ones headed southeast, while the "slow" ones headed northwest. The size of these directional differences was closely in line with predictions based on the changing azimuth of the sun: very strong evidence that they were employing a time-compensated sun compass.

Steve Reppert and his colleagues at the University of Massachusetts Medical School have since conducted a series of experiments that show how the monarch responds not only to the position of the sun in the sky, but also, like the honeybee and desert ant, to the e-vectors produced by light polarisation.[9] To adjust for the sun's changing azimuth during the course of the day, the butterfly—like the desert ant and honeybee—needs some kind of clock. This mechanism seems to rely on input from the antennae, since the animal loses its ability to time-compensate if they are chopped off or painted over, though it is not yet clear exactly how it works.[10]

Stanley Heinze and Reppert have found cells in the central complex of the monarch's brain that are tuned to specific e-vector angles, very like those earlier found in locusts. It is therefore possible that monarchs may be able to use e-vector patterns for orientation, even when the sun's disk itself is concealed by clouds. Since the e-vector pattern is potentially ambiguous, this means that in addition to tracking the sun's azimuth, the butterflies also need to measure its changing *height* in the sky. That process probably requires input from a second clock in its brain, though once again its nature remains to be established.[11]

What I have so far described is already an extraordinarily complex and sophisticated system, but there may be yet another dimension to it. As Perez suspected, it is possible that the monarch is also a magnetic navigator.

Patrick Guerra and Reppert[12] have carried out flight-simulator trials during which they exposed the monarch to artificial magnetic fields under a diffuse light. Though these involved only a small number of butterflies, the results suggested that the monarch may have an inclination compass. Guerra believes that it is based on light-sensitive receptors in the butterfly's antennae and that it acts as a backup mechanism when directional cues from the sky are unavailable.[13]

But not everyone is convinced. Mouritsen and Frost, who tested as many as 140 butterflies in their flight simulators, found no evidence of any magnetic orientation.[14] In a later displacement study, they tested the average flight direction of migratory monarchs—first in Ontario and then in Calgary, after being transported 1,500 miles to the west.[15] In Ontario, the butterflies generally headed in the right (southwesterly) direction for Mexico, as in the earlier study. In Calgary, too, they followed a similar course, which would eventually have led them to the Pacific Ocean, supposing they could have crossed the Rockies. So they did not appear to be capable of correcting for their westward displacement.

Mouritsen and Frost have also carefully examined a large volume of data from tagged butterflies recaptured over the years. Their conclusion is that the butterflies simply follow a general southwesterly course governed by their sun compass. However, another factor seems to be at work. Landscape features, like the high ramparts of the Rocky Mountains (which the butterflies cannot surmount), and the coast of the Gulf of Mexico (which they tend to follow as they prefer not to cross open water), act as physical barriers, effectively funneling the butterflies steadily southward toward Texas and then Mexico.

One last, major puzzle remains. While the various mechanisms I have described may well enable the butterflies to get within a few hundred miles of their final destination, it is still unclear how they pinpoint their overwintering sites in the mountains of central Mexico. One possibility is that in the closing stages of their journey, the butterflies home in on some kind of olfactory beacon, perhaps even the smell of the corpses of their dead brethren that cover the ground in their highland refuges.

The annual migration of the North American monarch is one of the most remarkable of all natural wonders, but it is one that future generations may never have the chance to witness. Not only are the forests in which the insects overwinter shrinking, as a result of illegal logging, but there are many other threats to the butterflies—including the profligate use of insecticides and herbicides that are either killing them directly or destroying the food-plants on which they depend. So time may be running short for scientists to fill in the last piece of this extraordinary puzzle.

∽

The people of the Maldive Islands, in the western Indian Ocean, have learned to expect dragonflies to appear among them in October. The most common of these insects is known simply as "The October Flyer" (Pantala flavescens) *and its appearance heralds the arrival of the northeast monsoon season. But where does it come from?*

Charles Anderson, who has studied the phenomenon closely, believes that most of these dragonflies (which are only two inches long) come from southern India or Sri Lanka and that they treat the Maldives only as a stopover. In fact, their ultimate destination seems to be East Africa, where the seasonal rains provide ideal conditions for their offspring. It is even possible that their descendants may carry on into southern Africa.[16] It is known that these insects can cover as much as 2,500 miles over land, but it now looks as if they can fly at least 2,200 miles over the ocean.

How is it possible for an insect—even such a strong flyer—to cover such enormous distances? The answer seems to be that they take advantage of high-altitude winds associated with the monsoons to give them a boost, and that they feed on smaller insects that are carried along in the same plume of fast-moving air. The likelihood is that millions of dragonflies make the journey and that, after breeding in various parts of Africa, their descendants return to India before the cycle starts again. In that case, the total round-trip journey could amount to as much as 11,000 miles. That would put even the monarch's more than 3,700-mile circuit in the shade—especially bearing in mind that the dragonflies, unlike the monarch, have to make long ocean crossings.

More recent research, based on measuring deuterium levels in the water found in the bodies of the dragonflies, bears out Anderson's hypothesis. In fact, it suggests that the dragonflies arriving in the Maldives have traveled even farther than he thought; they may have started their journeys in northern India or Nepal, or perhaps even beyond the Himalayas.

Though the October Flyer seems to be in a class of its own, flying insects are remarkably efficient migrants. If distance is scaled to body size, the longest insect migrations are approximately twenty-five times longer than those of the largest birds. One of the reasons is that insects are so adept at making use of the wind.

Chapter 16

THE SILVER "Y"

Many of the moths and butterflies that appear in Europe during the summer months will have made long migratory journeys to get there. Those that spend the winter in warmer latitudes come north to benefit from better supplies of food, and to avoid both predators and disease. The painted lady butterfly is a good example. Millions of these insects leave North Africa in the spring and, after several generations, their offspring eventually reach Britain, where they often breed in huge numbers. Their progeny then head south to avoid the northern winter. This is almost as long a journey as that of the monarch, and it appears that the painted lady too makes use of a sun compass.[1]

Another very impressive, though less colorful, migrant is the silver "Y" moth (so named because it sports white markings in the shape of the letter "Y" on its fore wings). These often turned up in the trap at my school, which is no surprise, since up to 240 million of them are estimated to reach Britain in a good year from the shores of the Mediterranean, where they spend the winter months.[2] After breeding, around three times that number may migrate south in the autumn. Since they are a serious agricultural pest, these moths have attracted a good deal of scientific attention, notably from Jason Chapman, who is a leading expert on insect migration based at the University of Exeter campus in Falmouth, Cornwall.

I traveled to Falmouth to interview Chapman. As a young boy, he spent all his spare time in the countryside around his home in South Wales, where he watched birds, as well as catching moths and butterflies. Like me, Chapman also bred caterpillars at his home. The books of Gerald Durrell and David Attenborough's TV films were a great inspiration to him, but Alfred Russell Wallace is his greatest scientific hero:

> What really interests me about Wallace is that—unlike Darwin— he was a totally self-made man. He didn't have a huge fortune, he didn't have a particularly good education and yet he did what he did. He went to the Amazon—his idea was to fund his own research by collecting and selling specimens. Most men would have been crushed by what happened to him when he came back. As he was sailing home, his ship caught fire and he lost everything. He got into a lifeboat and had to leave all his specimens behind. His life work went up in flames and then he almost died before being rescued. And yet, he did it all again, traveling for years through the rainforests of Southeast Asia.

Although no one in his family had ever gone to college, and his parents were not sure whether he could make a living as an academic, Chapman knew he wanted to make a career as a biologist. He studied at Swansea University, where his undergraduate project was all about how butterflies respond to sunshine. After gaining his doctorate at Southampton University, he became interested in insect migration and got a job at Rothamsted Research Station in Hertfordshire, where he began working with a device called Vertical-Looking Radar (VLR).

Using the reflections from the narrow beam of the radar that, as you may have guessed, points vertically upward into the sky, Chapman can not only spot individual flying insects up to a height of about 1,000 meters, but also determine their size, speed, direction, altitude, and in some cases even their species. With the help of this device, he has revealed the truly astonishing scale of the nocturnal movements of insects over southern England. Chapman estimates that *trillions* of them migrate annually from north to south and back again, and that their total mass amounts to several thousand tons.[3] Many of these migrants are silver "Y"s.

Chapman told me that when silver "Y" moths emerge from their pupae, they are primed to migrate as soon as possible. Their navigational system is simple. They have a preferred migratory direction (north in the spring and south in the autumn) and are programmed to fly for a certain period of time:

> They are in a totally migratory frame of mind in the first nights following emergence from the pupae, but during the process of migration their reproductive organs begin to mature. Over the course of maybe two or three days and nights, hormones are released which promote sexual maturation and then, when they are sexually mature, they stop migrating.

At this point, the males seek out the females and mate with them, and the females in their turn find food plants on which to lay their eggs. Whether the moths will have arrived in a place where their offspring can flourish depends on a number of factors, but the most important is the wind. The moths have to travel a long way—perhaps five hundred miles or more—in a matter of a few days, and if they had to rely on only their flight muscles, they would probably not get far enough. But with a strong wind under them, they can achieve a speed over the ground of as much as 55 miles per hour: if they can sustain that rate they can cover 350 miles or more in the course of a single summer's night. This rate of progress exceeds that of many migratory birds.

The newly emerged moths take to the air at dusk and appear to sample the air currents aloft. If the winds are heading in a broadly helpful direction, they commit themselves to the big journey. If not, they drop back down again to wait for more favorable conditions. They have only a few nights before their window of opportunity closes, and bearing in mind the British climate, millions of them must sometimes perish; but plainly enough of them survive to keep the race going.

Once the moths get aloft, they look for jets of warm, fast-moving air that will give them a strong boost. On a favorable night, every migrating moth appears to follow the same heading—to within a degree or two—over considerable distances, but they do not simply go with the flow. If the airstream is not going in quite the right direction, they make a course adjustment that will keep them closer to their preferred heading, and they can do this even when there is no moon and the stars are obscured by cloud.

Chapman's working assumption was that the moths probably had some kind of compass that enabled them to set their course. But, as we have seen, a compass cannot tell them if they are drifting sideways. The moths might be able to detect any cross-track error by observing landmarks, or the "optic flow" of the ground passing beneath them—if there was enough light. But Chapman thinks there must be times when it is simply too dark, or the moths are flying too high. It was a big puzzle.

An atmospheric physicist named Andy Reynolds, a colleague of Chapman's at Rothamsted, then came to the rescue. He did some mathematical modeling, which showed that the small-scale turbulence generated in a fast-moving air stream would be felt more strongly in the direction of flow than in other directions. If the moth could detect it, then it would be able to tell whether it was heading directly downwind. By comparing its compass heading with the wind direction, it could in principle determine whether it was drifting sideways, and it could then make the appropriate course correction.

That was interesting, but so far it was just a theory. Reynolds now made a prediction that could actually be tested. These "micro-turbulence" cues would, he calculated, be slightly displaced to the right (in the northern hemisphere) by the Coriolis force (*see page 186*). So if a moth were using them to identify the wind direction, it too would tend to show a slight right-hand bias. And that is exactly what Chapman has found. Here then was evidence that the moths could determine the direction of the airstream in which they were flying.

Chapman is certain that the silver "Y" has a compass sense that enables it both to set its initial course and later to correct it, when a cross-wind threatens to deflect it too far away from its preferred migratory direction. He suspects that it may well rely partly on the sun, but since the moths remain well-oriented throughout the night—even in the absence of the moon or stars—and can still make appropriate course corrections, that cannot be the whole story.

Chapman believes the silver "Y" must also have a magnetic compass at its disposal, which it may calibrate using skylight cues around sunset or dawn. But we must look elsewhere for hard evidence that a moth or butterfly uses the geomagnetic field to navigate.[4]

∞

The ancient murrelet is a perky little black and white seabird—a member of the auk family—that lives around the edges of the North Pacific. There is a large breeding colony on the remote islands of Haida Gwaii, off the coast of British Columbia.

When scientists tracked some of these birds to find out where they spent the winter months, they got a big surprise. Although only four birds returned safely to their burrows, it turned out they had traveled 5,000 miles across the Pacific to the waters off China, Korea, and Japan—and back again: a 10,000-mile round-trip that brought them back to a very precise location. The shortest route from Haida Gwaii would take the birds far north through the Bering Sea and Sea of Okhotsk, and the tracking suggested that the birds had indeed traveled that way.

No other bird is known to undertake a similar east-west migration in the Pacific, and why the murrelet does so is a mystery, as indeed is its method of navigation. The researchers think this extraordinary journey may reflect the route the birds took—in the long distant past—as they expanded their range from an original base in East Asia to North America.[5]

Chapter 17

THE DARK LORD OF THE SNOWY MOUNTAINS

When I visited Henrik Mouritsen in his office—housed in an old farmhouse with exposed wooden beams on the edge of the campus at Oldenburg in Germany—one of the many subjects we discussed was the research that he and Barrie Frost had carried out on the monarch butterfly. In the course of our conversation, he let drop that he was soon going to be traveling to Australia to take part in an investigation of the migratory behavior of another lepidopteron: the bogong moth.

This was an opportunity not to be missed, so I quickly asked if there was any chance I could join him. Mouritsen explained that the man in charge of the project was actually Eric Warrant and he kindly passed my request on to him. Things then moved quickly. Only a few weeks later, I went to visit Warrant in Sweden and, although we had only just met, he generously agreed to let me come along as an observer. And so, a month later, I found myself driving up into the Snowy Mountains at the tail end of the Australian summer. Having only a vague idea what lay ahead, I was both excited and slightly apprehensive.

Like the monarch, painted lady, and silver "Y," the bogong moth is a long-distance migrant. It breeds in southern Queensland during the winter months and then, to avoid the murderous summer heat, its newly emerged offspring head south in the spring to the Snowy

Mountains of New South Wales—a distance of more than 600 miles.[1] It has been estimated that two billion of them make the journey each year.

Canberra lies on their flight path and, attracted by the bright lights of the city, the moths have sometimes caused problems by plugging up lift shafts and ventilation ducts. At the opening of the Olympic Games in Sydney, an errant bogong made an unexpected appearance on TV, when it settled in the cleavage of an opera singer who was singing the national anthem. They are, according to Eric Warrant, esteemed and vilified in equal measure in their home country.

The ancient, heavily glaciated Snowy Mountains rise to heights of more than 2,000 meters, and on their summits stand piles of huge, weathered granite boulders, like the tors (high rocky hills) of England's Dartmoor, but on a far grander scale. The moths gather in the narrow fissures between these rocks, literally tiling the walls of the cool, dark crannies with their little bodies, up to 17,000 of them to each square meter of bare rock.[2] There they sit out the summer in a dormant condition known as aestivation: the summer equivalent of hibernation. If they are lucky enough not to be eaten by predators, they take to the air again in the autumn, and head north to renew the whole extraordinary cycle.

In two important respects, the bogong moth's achievement is even more remarkable than that of the monarch. For one thing, it flies only at night, while the monarch travels by day, so it cannot make use of a sun compass to maintain a straight course. The other big difference is that each moth (so long as it survives) is destined to make a complete round-trip of well over 1,200 miles—first flying south to the mountains, and then retracing its route to breed and then die in southern Queensland.

According to Stanley Heinze and Eric Warrant, who have written an entertaining account of this extraordinary moth's life history, if the monarch butterfly counts as the king of insect migration, then the bogong is certainly its "Dark Lord."[3] They sum up the navigational challenges it faces like this:

Bogong moths pinpoint a tiny mountain cave from over a thousand kilometers [more than 600 miles] away, crossing terrain they have never crossed previously, and locating a place they have never been to before. Moreover, they do all this at night, fuelled by a few drops of nectar and using a brain the size of a grain of rice. Don't even ask an engineer if they could build a robot equivalent! To achieve this remarkable behavior, the moth brain has to integrate sensory information from multiple sources and compute its current heading relative to an internal compass. It then has to compare that heading to its desired migratory direction and translate any mismatch into compensatory steering commands, while maintaining stable flight in very dim light while buffeted by cold turbulent winds.[4]

The bogong provides an ideal vehicle for exploring many of the questions that lie at the heart of animal navigation. Warrant's initial hypothesis was that the moth—like the dung beetle—was engaging in some form of celestial navigation. But unlike the beetle, which travels only a few meters, the moth flies all night long, and it may take several days or even weeks to reach its goal, depending on the winds. So whatever signposts it uses must remain reasonably stable. Polaris would meet that requirement, but it is invisible south of the equator, and since the moon, the Milky Way, and the stars are all in constant motion, Warrant could not see how any of them might supply the bogong with the information it needed:

I thought, God this is hopeless, they can't be using these cues, especially because in one of these experiments we actually blocked out the sky with a black cloth and the buggers still kept on going. Then it suddenly clicked—it must be the magnetic field. It was a big "aha!" moment. Birds have exactly the same job to do when they fly at night. In the northern hemisphere, they can use the rotational patterns round the Pole Star, but they also rely very heavily on a magnetic compass. What the hell, why not? Why wouldn't the moths be doing the same thing?

The road south from Canberra climbed slowly through sheep-farming country, which looked as if it had not seen rain for a long time. The roadsides were littered with the bloated corpses of careless kangaroos and wombats. Eventually I reached the small country town of Cooma. From there I headed toward the Kosciuszco National Park—the heart of the Snowy Mountains—and the landscape gradually emptied. The trees thinned out and there were fewer and fewer houses. This region was once plagued by bushwhackers: roving gangs of outlaws who terrorized the farmers that settled here in the early nineteenth century.

Warrant's house stands on a hillside, surrounded by snow gum trees, at the end of a long dusty track about nine miles from the nearest small town. Eric introduced me to the rest of the team: Barrie Frost; David Dreyer, and David Szakal from Lund; and Anja Günther from Oldenburg. Henrik Mouritsen was to join them after I had left.

The experiment I witnessed over the next few nights was a continuation of work they had begun several years earlier. The aim was simply to find out whether the moths rely on magnetic cues to find their way. The plan was to catch bogong moths at the start of their autumnal northward migration and fly them in a cylindrical arena, like the ones that Barrie Frost and Henrik Mouritsen had used in their earlier work on the monarch. Using a precisely calibrated coil system, they would then be exposed to various altered magnetic fields, and we would record their responses.

Where the bogong sleeps

When I arrived, the team had already been at work for some time and were running short of moths, so we needed to catch some more. Since we could not set up the light trap until it was dark, we decided to pay a visit during the day to the rocky crevices on the mountaintops where the moths gather in such vast numbers.

Eric and I, together with Anja Günther and David Szakal, made an early start for Thredbo, a ski resort in the steep valley of the Crackenback River. Since it was late summer, the town was very quiet, but we were able to take a ski lift up to an altitude of about 2,000 meters, and we then climbed through thick undergrowth and sphagnum bogs to reach the bleakly beautiful mountain tops. The moorland was dotted with flowers, and we soon found ourselves completely alone—apart from a few wild ponies and the ravens circling in the sky above us.

The Snowy Mountains are exceedingly old, and they look it. On every rounded summit, the tors rise in colossal, rounded sculptural shapes. Not many people know how to find the caves where the moths roost, but Eric led us to one of the best sites. There were few visible tracks to follow and at several points—ironically enough—we had to stop to check where we were. After a long hike under a fierce sun, we reached our destination: a towering mass of cracked and tumbled rocks at the top of a steep, grassy slope.

We clambered over some boulders to reach the mouth of one of the rocky fissures. A strong, nutty scent filled the air, and down at our feet, the ground was thickly carpeted with the disintegrating bodies of dead moths that had been sluiced out of their shelter by rainstorms. This was the source of the smell.

The gaps between the rocks were narrow, but we could just squeeze through. The air inside the crevice was filled with a fine dust of scales from the wings of the moths, which sparkled when it drifted across a beam of sunlight. Many of the moths had already left and a few were on the wing around us. With a flashlight, we could see patches where those that remained, their dun-colored wings folded neatly over their sleeping bodies, formed a perfectly regular pattern on the cold rock walls. They have no eyelids, of course, but the body of each moth acts as an eyeshade for the one behind it, so only the eyes of those in the foremost row are exposed to any direct light. It was the picture of tranquility and a testament to the efficiency of insect navigation.

Warrant explained that in the old days, before they were driven away by the colonists, the Aboriginal peoples from either side of this mountain range used to spend the summer months up among these rocky outcrops. They went there to escape the heat of the lowlands and to feast on roasted bogong moths, which apparently taste very good. It was a time for song and dance and the taking of wives. The early settlers recorded how the Aboriginal peoples were in much better condition as they returned from these moth-fueled festivities, their "skin being glossy and most of them quite fat."[5] But they are long gone, and their *corroborrees* are now only a distant memory.

There is some evidence that each cave is occupied by moths from one specific geographic location, though this theory remains to be confirmed. If so, the precision of their navigation exceeds that of the overwintering monarchs in the highland forests of Mexico, but even if the homing bogong is not quite that choosy, it still has to find a suitable cave, and that cannot be at all easy. Olfactory cues—perhaps even the nutty smell we noticed—may attract them.

Warrant's colleagues at Lund have recorded nerve signals from the antennae of bogong moths while puffing different aromas collected from the caves across them, but they have not yet managed to elicit any response. But since the moths they tested were taken from aestivation, it is possible that they were no longer motivated to react to them. Whatever the cues turn out to be, the southbound moths cannot have learned to recognize them, because they are all novice migrants. Their attraction to it must be instinctual. Here then are some fascinating questions that remain to be answered.

By the time we started our descent, the sun was already sinking, and it was getting dark when we reached the point where we set up the light trap. Though not very sophisticated, it was effective. It consisted of a large, powerful floodlight powered by a portable generator, and a white sheet stretched between a couple of scrubby trees. Within a minute or two, it was attracting all manner of insects, most of which were not bogongs. One of them was an enormous hairy cicada that Eric found fascinating.

As an insect-lover, I was entranced by the spectacle of so many unfamiliar flying insects, but identifying the bogongs was not easy for a beginner like me. I also had great difficulty in catching them—unlike the two youngest members of our party, whose reactions were a great deal faster than mine.

Next morning, we had the task of "stalking" the moths ready for the experiment. This is a crucial part of the ingenious tethering process. The moths are first chilled in a portable icebox to make them sleepy, and then gently immobilized under a piece of wire mesh held down with weights. The next step is to strip the furry scales from a small area of the thorax (the middle section of the body just behind the moth's head), with the help of a miniature vacuum cleaner powered by a car's electric fuel pump, which Barrie Frost had improvised.

The exposed carapace is now ready to receive a tiny dollop of adhesive, and a length of thin tungsten wire with a tiny loop on the end is quickly glued to it. It is essential that this stalk is vertically aligned, otherwise the flying moths cannot maintain a constant heading. Once successfully "stalked," the moths are put separately in small boxes, each supplied with food in the form of a Q-tip loaded with honey, and kept cool and dark until their services are needed. By the time the stalk is in place, the moths are usually waking up, and they sometimes escape while being transferred to the boxes. Catching them again is not easy.

The site of the experiment was the hilltop above the house. A power cable had been laid and a small tent set up to shelter the recording equipment and control apparatus for the magnetic coil systems, as well as the people operating them. Around sunset, we plodded up the hill, avoiding the large piles of kangaroo dung, while carrying the moths in their cooler and all the other gear—including tea and biscuits. The temperature dropped fast, and during the night the thermal underwear Eric had lent me was very welcome.

There were two cylindrical arenas (like those in which Mouritsen and Frost had tested the navigational skills of the monarch), and across the top of each, a Perspex arm carried an axle to which the stalk of each moth could be attached. The moths were then free to "fly" in any direction they chose. A moving pattern projected onto the floor of the cylinder created "optic flow" that encouraged them to take off, and a feedback system ensured that the flow was aligned with the direction in which they flew.

The heading chosen by the moth was monitored electronically and relayed to laptop computers in the nearby tent. Using the coil system around the arena, it was possible to rotate the magnetic field by a precise amount, and then see exactly how the moth reacted to the changes.

If at first you don't succeed . . .

When Warrant and his team first tried this experiment, it was a complete failure. The moths generally did not respond at all to the changed magnetic fields, though occasionally—and inconsistently—there was a big effect. After three frustrating years, they were beginning to think that the moths either did not have a magnetic compass, or that it was impossible to grasp how it worked. Then it suddenly occurred to Warrant that the moths might be responding to visual, as well as magnetic cues:

> The thing is, we've got this bloody arm across the top of the arena, we've got the coils visible, and moreover the wall of the arena, which was lined with cardboard, started to buckle after a few dewy evenings. Even though it's almost invisible to us, I know enough about insect vision at night—which is superb—to know that the moths can see all of this. And I'm thinking, we're dickheads. They can see all of this and they're using it.

What were they to do? It was impossible to eliminate all possible sources of visual information, so they installed a small, horizontal diffuser disk on the axle just above the stalk, to prevent the moth from seeing anything above it. The disk would, however, allow the faint ultraviolet light from the night sky to reach the moth. This was essential, as it seemed likely that the animal's magnetic-compass sense depended on it. But there was still the problem of the arena wall with all its imperfections.

Warrant came up with a neat solution:

We decided to install some really strong landmarks that would override the subtle ones. The side was kind of pale grey to start with, so we put in a black horizon and then we introduced mountains—just black triangles on a piece of clear film that we could flip in and out, so that they sat on the horizon either at zero degrees [due north] or at 120 degrees [roughly southeast by east].

Now at last they started to get some useful results:

We then did a four-phase experiment, each of five minutes—a total of twenty minutes. First phase, Earth-strength magnetic field in its normal northerly alignment at zero degrees, with a mountain also at zero degrees, so everything was in the same direction. Then after five minutes of flight, we changed everything to 120 degrees—mountain and field again pointing in the same direction. Around the moths went, not all of them, but enough to show a definite effect. In phase three, we left the mountain where it was and returned the field to zero.

Then all hell breaks loose! They keep flying for two minutes toward the mountain and then they lose it and become completely disoriented. In the fourth phase—the final five minutes—we put the mountain back to zero degrees and the moths find their way again. But during the third phase—with the cue conflict—they get into real trouble. It's really clear from the data that we have a real effect.

The fact that we can cause this confusion with the magnetic field change means that they have a magnetic sense. If they didn't and they were just going for the mountain, they'd have done that in the third phase and been perfectly oriented. And what makes this all the more impressive is that we were four meters away, and just pressed a button to change the field, so we weren't physically interfering with the moths at all.

This initial "cue conflict" experiment convinced Warrant that the moths were doing exactly what a human helmsman does when steering a compass course at sea. Rather than staring at the compass card all the time, sailors find it easier to get the vessel on course and then line up the bows with a distant cloud, or perhaps a star, and then steer by that. Occasionally, they will look back at the compass to check that they are still on the right heading. The moths also seem to be setting their course, initially by reference to their magnetic compass, and then using whatever visual cues are available to them (in this case the "mountains" inside the arena) to stay on track.

It would be understandable if they got confused when the magnetic field around them suddenly changed. Should they then stay with the visual "landmarks," or adjust their course in keeping with the magnetic signal? Warrant believes the magnetic compass dominates the landmarks, and that the delays occur because the moths check their course against their internal compass on average once every two minutes. This system has a great advantage over a solar or lunar compass: It does not call for any kind of time compensation.

Of course, none of this is easy to demonstrate with complete scientific rigor. The data are always noisy because the moths do not all behave in exactly the same way. This may be partly a result of genuine individual differences among the moths, but other influences—like poorly attached stalks or distracting light or sound—may also be to blame.

So when I joined Warrant's team, the job facing them was to carry out a fresh batch of experiments, to eliminate all possible confounding factors. In particular, they needed to randomize the order in which the different cues were presented to the moths, rather than always starting with everything pointing in the normal, due-north migratory direction, as had been done the year before. After nightfall, the sky above us was a magnificent sight. Even out in mid-ocean, far from any light pollution, I have never seen so many stars. The Milky Way glowed brilliantly. I could make out the dark patches of dust within it that you can usually see only in long-exposure photographs. The Southern Cross rose majestically in the southeast and, near the southern celestial pole, in an empty patch of sky, the two Magellanic Clouds—our nearest galactic neighbors—stood out clearly.

Inside the tent, we sat up until the early hours testing twenty or thirty moths each night. The procedure was carefully standardized, and we tried hard to avoid showing any light or making any sound near the arenas. Each test started with a period during which the moths were allowed simply to establish a preferred heading in the natural geomagnetic field. Then they were exposed to the four different test conditions in a preset, randomized sequence.

Four of us sat close together in the tent on folding chairs, watching the two laptops that recorded what the moths were doing. We called out when it was time for our colleagues to change the magnetic field or move the "mountains." We could see exactly what each moth was doing after it was put in the apparatus: Sometimes they settled quickly down to fly in one direction—often but by no means always northward—but other times they went whizzing all around the compass. This problem seemed to be caused by the incorrect attachment of the stalk. Once they were settled, Eric, sitting on his own at the back of the tent, turned on the two coil systems and we watched what happened.

At the start, it looked as if a lot of the moths were "misbehaving," though gradually a pattern began to emerge. The temptation to exclude results that do not fit the theory is strong—and not all scientists succeed in resisting it. By massaging the data, you can obtain results that appear "statistically significant," though they are, in reality, totally misleading, so it is vital that all the valid data should be included.

Experiments like this call for a great deal of patience, and jokes—even bad ones—lighten the mood: Eric's surprising admiration for the big, hairy cicada we had seen at the light trap was a running gag, with unexpected comic potential. It was a relief when the supply of moths finally ran out, and we could stumble back down the dark hillside and have a shot of whiskey before taking to our beds.

The experiment continued for several more weeks after I had left, and it was some months before the results of the experiment were fully analyzed. All those nights on the cold hilltop in New South Wales have certainly paid off. The use of a magnetic compass has at last been convincingly demonstrated in a flying insect. More than that, an entirely new navigational strategy—which involves comparing both visual and magnetic "snapshots"—has been uncovered. Something never before seen in any animal.[6]

There used to be tales of baby pet alligators flushed down toilets in New York surviving to form colonies in the warm underworld of the city's sewers. That does not sound very plausible, but in southern Florida, escaped exotic pets have become a real pest. Burmese pythons—among the largest snakes in the world—have in recent years made a home for themselves in the subtropical wetlands of the Everglades, where they are making quite a dent in the local wildlife. They have also extended their range down into the Florida Keys.

One way of controlling the spread of invasive animals like these is to move them away from areas where they are causing trouble, but first you need to be sure they will then stay put—especially in view of the experience with Australian crocodiles (see page 66).

So scientists captured pythons in the Everglades, implanted radio track-ers in them (under anesthetic), and transported them in opaque, sealed containers to sites up to twenty-two miles away. Six of the snakes were released at these remote locations, while six others (the controls) were taken straight back to the places where they were captured, before being given their freedom.

The radio tags on the pythons were monitored from light aircraft. To everyone's surprise, the translocated pythons all headed for home, and five of them returned to within three miles of the places where they had been caught. They were more active and moved faster than the controls, and they clearly had a good idea of where they wanted to go. The controls, on the other hand, just moved around randomly.

It seems unlikely that the homing pythons used DR, so perhaps they have some kind of map, based on magnetic, olfactory, or celestial cues. Behavior like this has never been seen before in a snake.[7]

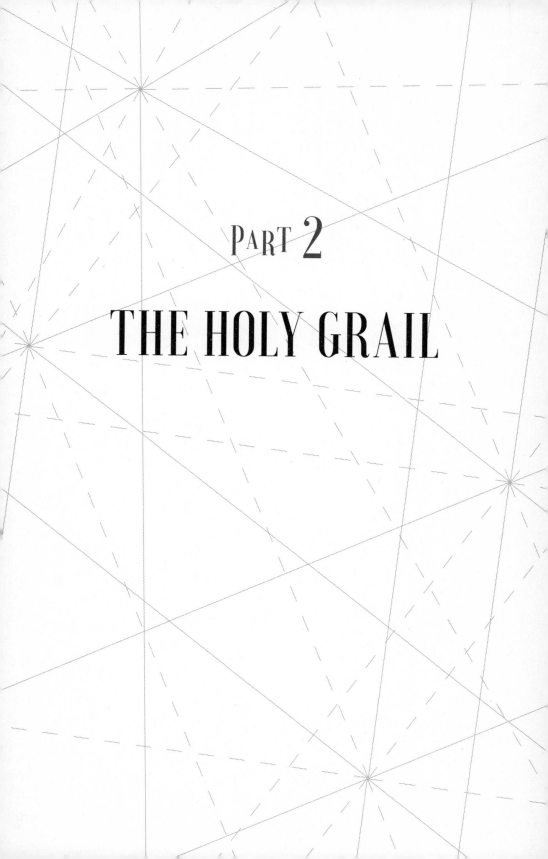

PART **2**

THE HOLY GRAIL

Chapter 18

MAP AND COMPASS NAVIGATION

In front of me is an old British Admiralty chart of the North Atlantic. Down the left-hand side runs the coast of North America—from Resolution Island at the mouth of the Hudson Strait to Jupiter Inlet on the coast of Florida. On the eastern side its limits are marked by two groups of islands: the Faroes in the far north and the Canaries in the south. On the top edge Cape Farewell, the southern tip of Greenland, pokes its nose in, too. But the chart is, of course, dominated by the vast expanse of ocean. Dotted with soundings showing the depth of the water, it also bears three compass roses, on which true north is marked with a purple star, a device that recalls the old name for Polaris: *stella maris*, the "star of the sea."

Charts like this may not look all that special, but an extraordinary amount of hard-won information is packed into them. In command of small sailing ships, and often working from open boats, the young naval officers who made them risked their lives and endured all kinds of hardships in surveying such remote and dangerous places as Alaska, Tierra del Fuego, or the malarial coasts of tropical Africa.

Tens of thousands of soundings and compass bearings had to be taken, and at every opportunity, positions had to be fixed accurately by reference to the sun, moon, and stars. It was a truly heroic endeavor. Nowadays electronic depth sounders, GPS, and satellite imaging have greatly simplified the task, but making a chart is still a very exacting process.

In the preface, I discussed briefly the different ways in which a visitor arriving in a strange city might learn to find their way around it—without using GPS. We saw that they could do so either with or without the help of a map. These two approaches are conceptually distinct and have been labeled by scientists as *other-centered navigation* and *self-centered*.[1]

When you are navigating in a self-centered fashion, all that matters is how objects in your environment relate to yourself. You take note of prominent buildings, you remember which way you turned at a key junction, and so on; but in every case the world revolves around *you*. We have already seen many examples of self-centered navigation at work—from the desert ant to the bogong moth.

At its simplest, self-centered navigation depends on learning to recognize the landmarks that define a route, so that you can accurately retrace your steps. Our imaginary tourist could thus get back to their hotel by following a sequence they had observed on their outward route, but in reverse order.

Then there is DR. Though a bit more complicated, it too is a form of self-centered navigation. It involves putting together information about the course you have followed and the distance you have traveled, so that you can always plot your position in relation to your starting point. Employing DR, our visitor would be able to maintain a constant sense of the direction in which their hotel lay and how far away it was—like one of Wehner's foraging ants. Rather than simply retracing their steps, they would then be able to find the most direct route back to their hotel.

These two forms of self-centered navigation are not mutually exclusive, and many animals, including humans, use both. But neither will work unless you are able to monitor your progress without any breaks. If you suddenly find yourself in an unfamiliar location with no idea how you got there, and cannot detect any signal that would help you find your homeward bearing, neither system will be of any use to you. In that situation you will either need a lot of luck or some quite different way of determining which way you need to go.

This is where *maps* come in, and that means switching to other-centered navigation.

Other-centered navigation depends on grasping how the objects that surround you are geometrically related to *each other*. Printed maps—like the chart of the North Atlantic—provide just this kind of information, as do the digital ones on which we now usually rely. They are based on a system of coordinates, of which latitude and longitude are the most familiar.

But a map will be of little use unless you have some way of fixing your position on it. One way of doing this is to match landmarks you can see with the symbols that represent them on the map. But that system will not work if you are far out at sea, or in the middle of a featureless desert, where there are no landmarks to consult. Unless you have some other way of determining where you are, you will be at a loss—if not actually lost.

We humans have a variety of tools for fixing our position without the help of landmarks, of which GPS is only the latest and most accurate. If you can read out your latitude and longitude on some gadget, it is a simple matter to plot your position on a chart. Then you can quickly work out the course that will take you to your chosen destination, wherever that may be, using a ruler and protractor.

So, for example, if you were in 40 degrees north and 40 degrees west, you would quickly discover that you were in the middle of the North Atlantic—about 420 nautical miles (483 miles) west of the island of Corvo, in the Azores. And if you wanted to head for New York, the chart would tell you that a course just a little north of true west would take you there.

The process I have described here is known—for obvious reasons— as *map and compass* navigation.[2] One of the deepest questions facing students of animal navigation is whether a system like that is within the reach of any nonhuman animals and, if so, how it works.

The central issue is whether animals can fix their positions when they find themselves in unfamiliar places without access to any recognizable landmarks, and whether they can then determine the course and distance to their goal. They obviously cannot make use of navigational satellites, but perhaps, like us, they have some way of working out where they are that relies on signals they pick up from distant sources. These might, for example, be sounds, smells, or features of the earth's magnetic field.

From a human perspective, this is a fairly bizarre idea, so one or two practical (if fanciful) examples may help.

Suppose you know that the aroma of hops comes from a particular brewery. You can work out which way you are facing by observing the *direction* from which the wind is carrying it: If you are facing into the wind, then the brewery must be ahead of you. And if—following a wind shift—you then detect the smell of lavender coming from a field that lies in a different direction, you can work out (very) roughly where you are on a mental "map" that marks the positions of the brewery and the lavender field. Since you are relying on directional information, this would count as a *vector map*.

But you might also be able to make use of changes in the *character* or *intensity* of the signals reaching you. Suppose you had a mental map on which the *loudness* of the sounds coming from three separate sources (perhaps a bell tower, a pile driver, and a rifle range) were plotted in the form of gradients. Concentric circles could relate the strength of each distinct sound to the distance from its source. By working out (somehow) where the circles matching the observed loudness of the three signals intersected, you could, in theory, obtain an approximate fix.[3] In the real world, the wind and other factors would make such a system very unreliable, but I hope the general idea is clear. *Gradient maps* of this kind could in principle be based on other signals, including olfactory ones.

Since local cues—like sounds or smells—do not normally travel very far, it is hard to see how an animal could use them to fix its position, unless their sources were reasonably nearby. But some cues—such

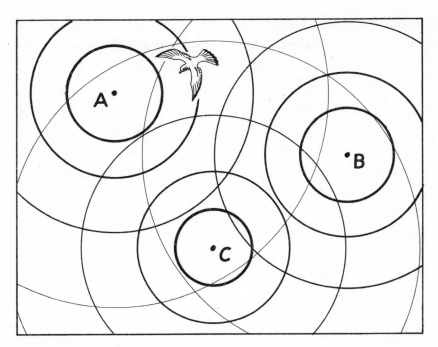

A hypothetical gradient map. A, B, and C here represent sources of distinctive sounds. The concentric circles show how they diminish in strength as they spread out.

as celestial or magnetic ones—are *globally available*, and some animals may use them to perform long-range map and compass navigation.

In theory, an animal could determine its position by making observations of the sun and stars, like a human navigator using a sextant. But it would need two clocks and detailed information about the precise movements of the heavenly bodies it was observing. That sounds like a pretty tall order, and there is no evidence that any animal can actually fix its position in this way. We certainly cannot do so without the help of technology.

Making use of the geomagnetic field would depend on measuring two or more of the parameters that define it, such as intensity and inclination, and knowing how they varied across the surface of the earth. The gradients could in principle supply an animal with a coordinate system, rather like latitude and longitude, that would allow it to plot its position on a magnetic map.

Nonhuman animals—like our imaginary tourist—could also acquire map-like representations of their worlds simply by exploring their surroundings. Though it is easier for us to grasp how such maps can be built up on the basis of visual information, they do not have to be. An animal might learn to associate different locations in its home range with unique smell or sound combinations. Each one of these would be like a small tile and, when put together, they would collectively form the basis of a *mosaic map*, which might help it work out where it was (at least approximately) without even opening its eyes. Plainly such a map would be of no use if the animal ventured into unfamiliar territory.

It is difficult to know what the scale or precision of these different kinds of maps might be. Much would depend on the sensory and cognitive capacities of the animal and the quality of the information available to it. And, of course, a variety of maps could be used in parallel. Perhaps, over the course of its long life, a wandering albatross might build up vector, gradient, and mosaic maps that embraced an entire ocean—based on a host of different cues. Coupled with a compass, these might provide the basis of an accurate, geographically extensive, long-range navigation system.

So much for theories. I now want to explore the evidence that nonhuman animals actually use maps, rather than relying on simpler, self-centered navigational techniques.

Perdeck's starlings

The story starts in the 1950s, when a Dutch scientist named Albert Christiaan Perdeck (1923–2009) carried out a long series of experiments—of a kind that would not now be allowed—in which thousands of starlings (both adults and juveniles) were caught and ringed near The Hague in the middle of their westward autumn migration. They were then transported by air to various sites in Switzerland, hundreds of miles away from their normal migratory route, and released.

Sometimes the adults and juveniles were mixed together and on other occasions they were separated. In normal circumstances, the ringed birds would all have headed west-southwest from The Hague toward their winter quarters in northwestern France, but not all the displaced birds maintained that course. Perdeck showed that the adults generally compensated for their "sideways" displacement and headed in a northwesterly direction. Most of the juveniles, when traveling on their own, continued flying in a southwesterly direction and ended up in the south of France or Spain. But when the juveniles traveled in company with adults they, too, followed a revised course. Perdeck also noticed something else: The displaced juveniles tended in later years faithfully to return to the "wrong" region—the one in which they had wintered after their first displacement—a region they would never otherwise have visited.[4]

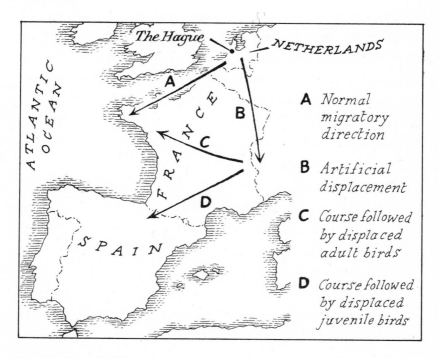

A *Normal migratory direction*

B *Artificial displacement*

C *Course followed by displaced adult birds*

D *Course followed by displaced juvenile birds*

Perdeck's starlings

Perdeck interpreted these results as evidence that adult starlings knew where they were going and had access to some kind of map, while the juveniles (when left to their own devices) merely followed a genetically programmed compass heading until, when the migratory urge left them, they simply stopped. Although he assumed that the *capacity* for "map and compass" navigation was innate, Perdeck argued that the birds could only employ it after they had visited their migratory goal at least once. In other words, instinct was not enough; the birds needed also to acquire some geographical knowledge on their journeys. This, he believed, explained the difference in performance between the adults and the first-time migrants.

The Perdeck study (which had the great merit of relying on the natural behavior of birds in the wild, rather than hopping around in Emlen funnels—*see page 172*), and others like it, has encouraged the belief that some birds are "map and compass" navigators. But this is a big claim and it is hard to exclude other, simpler, explanations. Perhaps the adult birds are genetically programmed to head off in the right general direction and fly for a certain length of time. Having once arrived at a congenial location, they might then learn to recognize some local beacon, perhaps olfactory or auditory, that would draw them back in future years, even from a long distance. Or they might simply be learning a sequence of landmarks along the route. Are they perhaps using celestial or magnetic cues—or is it some combination of the above?

Pigeons are the lab rats of the avian world and have been the subject of more research than any other birds. Some researchers argue that the extraordinary homing abilities of pigeons can only be explained on the basis that, in addition to a magnetic compass, they also have access to a map—one that is not based on visual information.

One of the most startling pieces of evidence for this claim comes from a series of experiments in which pigeons were fitted with frosted contact lenses that prevented them from identifying any landmarks. Even when displaced to distances as great as eighty miles, these birds could often find their way back to within a few miles of their home lofts, though they had much more difficulty in doing so than birds wearing clear lenses.[5] The puzzling fact that birds delivered to an unknown, distant release site under anesthesia (which rules out the possibility that they have learned the outward route or employed DR) can home successfully also demands an explanation.[6]

Assuming that pigeons do make use of smells to help them navigate, it is possible that they can follow a scent trail—like a moth. But this would only work if the pigeon happened to find itself downwind of its loft. So maybe they make use of olfactory maps of some kind. These might take the form of a learned pattern of smells that form a *mosaic* (though that would not explain how they can home from unfamiliar locations), or they might be based on *gradients*—for example, geographical variations in the relative strength of the individual smells that form characteristic "bouquets."[7]

The latter notion may sound far-fetched, but there is some evidence that, despite the effects of air turbulence, varying mixtures of chemical compounds are stably distributed over wide areas, and therefore could in principle support a gradient map of this kind.[8] But since nobody has yet demonstrated the navigational use by pigeons of any naturally occurring odor combination, the theory remains speculative.

Infrasound, too, might perhaps provide the basis of a gradient map, though Hagstrum's hypothesis is that the infrasound "signature" of the home loft area acts like a beacon, in which case there is no need to invoke the use of an "acoustic" map.

Pigeon racers have often reported that their birds are sensitive to solar storms that disrupt the geomagnetic field. They can also be disturbed by magnetic anomalies caused by local concentrations of magnetic material in the earth's crust. These observations have encouraged the notion that magnetic information may be important to them, and it has often been suggested that they have access to some kind of magnetic map. Such a map would probably have to be based on gradients in the geomagnetic field, but it is also possible that they might be able to make use of magnetic anomalies as simple landmarks.

But a magnetic gradient map based on magnetic intensity and inclination could not be very accurate, and it is hard to see how pigeons could use one to home successfully. This is a simple matter of physics. While both intensity and inclination show strong north-south gradients—and could therefore help a bird determine its latitude—in most parts of the world they vary only slightly from east to west.[9]

And that is not the only difficulty facing proponents of the magnetic map hypothesis. The daily variation in field intensity would completely swamp the very slight changes that the pigeons would need to detect in order to home within a radius of a few miles. Henrik Mouritsen put the problem to me like this:

> There is one very simple consideration. What's the magnetic field intensity at the magnetic north pole? About 60,000 nT. What is it at the magnetic equator? About half that—30,000 nT. So there's a difference of 30,000nT between the two. How far is it around the earth at the equator? About 25,000 miles. So the distance from the equator to the pole is about one quarter of that: 6,250 miles. Now how much does the magnetic field change on average per mile? By just 3 nT. But what is the daily variation? Thirty to one hundred nT.[10]

There remains the theoretical possibility that a pigeon could make effective navigational use of intensity gradients by averaging the signals over time, but it could only do so if it moved very slowly or stopped frequently—and that is not how these birds actually behave.

So an intensity/inclination magnetic map would simply not be accurate enough to enable a pigeon to home successfully.

But this does not mean that magnetic maps are of no use to other animals. Pinpointing a precise location is a very stringent navigational challenge, and some migratory birds—as well as animals like turtles, salmon, and lobsters—may be able to use magnetic maps for other, less demanding purposes.

We have already seen how important polarized sunlight is to insects, and there is also evidence that migratory birds may use it to help calibrate their sun compasses,[11] but it might also be of navigational value to marine animals.

More than fifty years ago, Talbot Waterman showed that e-vector patterns were visible underwater—even down to a depth of 200 meters. Their orientations are directly related to the position of the sun, and they can therefore be used to determine direction in much the same way as e-vectors in the sky.[12] It has long been recognized that underwater e-vectors could therefore provide the basis of a sun compass, but new research shows that they could also help an animal determine its position.[13]

Using a polarization sensor that mimicked the visual system of the mantis shrimp, scientists have shown that animals could, in principle, work out both the azimuth and the altitude of the sun—and thereby determine their approximate location. Recordings made at various locations around the world, at different depths and times of day, suggest that such a system could generate surprisingly accurate fixes, as well as compass headings.

Many marine animals—including salmon—are known to be sensitive to polarized light, but since this navigation system poses exactly the same problems as other forms of celestial position-fixing it is hard to believe that any marine animal actually employs it. However, we have been surprised before, so perhaps it would be best to keep an open mind.

Chapter 19

CAN BIRDS SOLVE THE LONGITUDE PROBLEM?

S cientists have been trying for a long time to work out what role, if any, maps play in bird navigation, but until recently the picture has remained very confused. The problem is a genuinely hard one, though the difficulty in generating consistent results may also reflect the fact so many different species have been studied: a starling, after all, is not much like a shearwater. But things are starting to change. Over the last ten years or so, a number of experiments have yielded compelling—if not yet decisive—evidence that some birds may indeed employ some form of map and compass navigation.

In 2007, Kasper Thorup published the results of a remarkable study[1] that provided the first really solid evidence that daytime migratory birds, in this case white-crowned sparrows, could somehow compensate for a large west-east displacement. It seemed they could detect a large change in longitude.

Thorup captured the sparrows (both adult and juvenile) when they were taking a break at a stopover site in Washington State, while on the way from their summer breeding grounds in Canada and Alaska to their overwintering sites in the southwestern US and Mexico. The birds were then transported by air eastward to Princeton in New Jersey, a distance of 2,300 miles, where tiny radio-trackers (weighing only half a gram) were glued to their backs.

After a day or two of rest, the birds were set free: the juveniles at one site, the adults at another, to avoid the risk that the youngsters would follow their elders. A total of thirty birds were tracked (fifteen adults and fifteen juveniles) with the help of observers in two light aircraft. Each bird's last stopover site was recorded, and these locations were then used to calculate its preferred migratory heading.

The normal migratory route followed by these birds is a southerly one, but the displaced adults consistently headed off in a westerly direction—as if to compensate for their unhelpful transcontinental journey. The inexperienced juveniles, on the other hand, headed south, as if completely unaware of the trick that had been played on them.

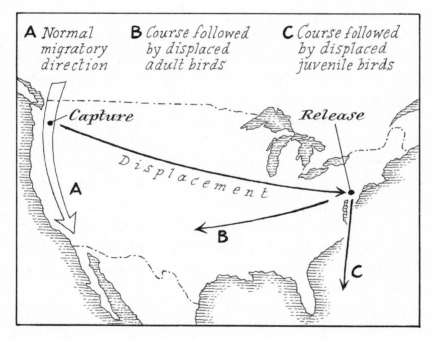

Thorup's sparrows

Thorup concluded that the adults must have acquired a "navigational map" that operated on a continental, or perhaps even a global scale. This enabled them to determine where they were, even after the massive longitudinal displacement, while the juveniles continued to rely on a simpler, innate directional program.

While suggesting that magnetic cues might form the basis of the sparrows' map sense, Thorup acknowledged that the differences in magnetic *intensity* between the west and east coasts of the US were too slight to be of any navigational value. He speculated that they might be employing celestial or olfactory cues, but ruled out the possibility that they could have tracked their changing position using some form of DR, on the grounds that the distance involved was too great.

Further evidence for the existence of a map sense in birds comes from a series of experiments conducted by two Russian scientists, Nikita Chernetsov and Dmitri Kishkinev, in collaboration with Mouritsen's group in Germany.

During their spring migration, reed warblers pass through Rybachy on the Baltic coast, heading for breeding grounds that lie far to the northeast. Chernetsov captured birds there and transported them (by aircraft) 600 miles due east to a point near Moscow. The birds therefore experienced no change in latitude of the kind that could be detected with an inclination or stellar compass. Had the birds been unaware of their easterly journey, they would presumably have still wanted to fly in a northeasterly direction. But when tested in Emlen funnels under clear, starry skies, the adults showed a strong desire to head toward the northwest—exactly the right direction to take them to their breeding grounds from the new location.[2] They seemed to know what had happened to them and corrected their course accordingly. The juveniles, however, were oriented in a northeasterly direction.

Chernetsov noted that there was a slight difference (3 percent) in magnetic intensity between Rybachy and the site to which the birds were displaced. It was therefore theoretically possible that they might have been able to use this cue to detect the change in longitude. But this seemed improbable.

Alternatively, perhaps the birds had used a difference in the times of sunrise and sunset at the two sites to work out the longitude difference. This would mean they had two internal clocks: one that kept

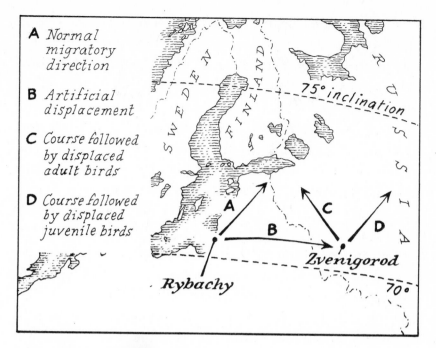

A *Normal migratory direction*

B *Artificial displacement*

C *Course followed by displaced adult birds*

D *Course followed by displaced juvenile birds*

75° inclination

SWEDEN

FINLAND

RUSSIA

A

B

C

D

Zvenigorod

Rybachy

70°

The warblers of Rybachy. Note that there is no difference in magnetic inclination between the two locations.

Rybachy time, while the other quickly adjusted to the solar time in the new location.

While there is no evidence that birds can make comparisons of this kind, the "circadian clock" in mammals (based in a part of the brain called the hypothalamus) does contain two types of neuron, one of which reacts immediately to a change in daylight hours, while the other takes up to six days to adjust.[3] These two clocks might just possibly enable mammals—and perhaps also birds—to detect a change in longitude.

To test this intriguing "double-clock" idea, Kishkinev conducted an experiment in which migratory warblers underwent an artificial clock-shift.[4] First, he tested warblers in an Emlen funnel to establish their preferred migratory direction in the usual way. Without moving them from Rybachy, he then gave them a mild case of "jet lag," by artificially altering the time of sunset and sunrise to match the regime at the site near Moscow. If the birds really were relying on a double-clock system to track changes in their longitude, the jet-lagged birds should have altered their preferred heading, but they failed to do so. This was strong evidence that the displaced birds must be using some other mechanism to work out where they were.

Were the birds tracking their easterly movement using some form of inertial DR? Were they using olfactory or auditory cues—or were they secretly engaged in some sophisticated form of celestial navigation?

Chernetsov and Kishkinev neatly disposed of all these possibilities by conducting an experiment in which the warblers were not physically moved at all. Instead, they simply surrounded them with an altered magnetic field that exactly matched the magnetic signature of the location 600 miles to the east.[5] Once again, the birds changed their preferred direction; in fact, the response was "indistinguishable from the one seen after a real physical [600 miles] eastward displacement." Since nothing else had changed, the only cues they could have been using were magnetic. But what exactly were they?

The same team has also shown that the warblers are unable to compensate for their eastward displacement if the trigeminal nerve connecting their upper beak to their brain is severed.[6] This suggested that "some sort of map information" was being transmitted to the brain via this channel, though what that might be and what sensory apparatus it came from remained unclear.

Magnetic declination

If measurements of magnetic intensity and inclination do not offer much useful information about changes in longitude, perhaps magnetic declination is the key.

Declination, you may recall, is the angular difference between true north and magnetic north, and it varies widely across the surface of the earth. Chernetsov and his colleagues have now tested whether a change in magnetic declination affects the behavior of warblers during their west-southwesterly autumn migration. In doing so, they have made a very intriguing discovery.[7]

This time they exposed both adult and juvenile birds to an altered magnetic field that matched the one found at Rybachy in all but one respect: The declination was rotated counterclockwise by 8.5 degrees. The altered field closely matched that found near the Scottish town of Dundee, which is roughly 900 miles to the west and far removed from their normal migratory route. All the other information available to the birds—magnetic intensity and inclination, olfactory, celestial, and auditory—necessarily remained unchanged and would have told them they were still in Rybachy.

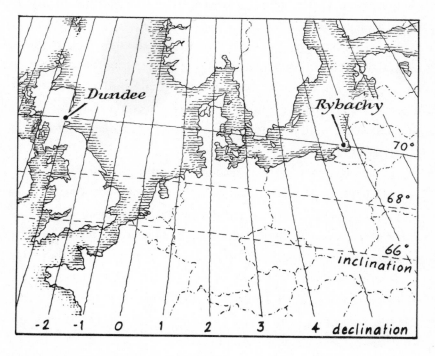

Migratory warblers may be able to determine their longitude by measuring changes in magnetic declination.

The results were exciting. When tested in Emlen funnels under moonless, starry skies, the adult birds responded with a "dramatic 151-degree change in their mean orientation" from west-southwest to east-southeast—a course that would have taken them toward their intended goal if they really had been in Dundee. By contrast, juvenile birds exposed to the same declination change did not alter their orientation; they just got confused.

In order to alter their migratory heading in response to changes in magnetic declination, the warblers would need to have tracked the difference between the bearings of true and magnetic north. But how is that possible? The best guess is that they establish where true north is by inspecting the rotating pattern of circumpolar stars, and then comparing that with the information coming from their magnetic inclination compass.

In line with Thorup's observations—and the much earlier work of Perdeck—the new study suggests that the experienced, older birds have acquired information about their normal migratory route that is not available to the juveniles. The ability to compensate for longitude changes would therefore be a learned, rather than a hardwired, inherited skill.

Mouritsen acknowledges that the Emlen funnel is a highly artificial environment, but he points out that the experimenter does at least know exactly what is going on inside it. The inputs can be controlled and one factor can be altered at a time. Mouritsen has tested birds by tossing them in the opposite direction to the one in which they have been jumping during the experiment and watching where they go. They normally head back in the "right" direction. And he says that the results from Emlen funnel tests match fairly consistently the observed behavior of free-flying birds.

Anna Gagliardo, however, has her doubts. In the old days, the navigational ability of pigeons was often assessed by watching them through binoculars until they disappeared from view. Sometimes birds that were heading homeward at this point failed to return to their lofts, and conversely, some birds that were wrongly oriented reached home successfully. Gagliardo therefore thinks that testing birds in Emlen funnels is not a reliable way of determining their real navigational preferences.

There is another issue. Since the declination difference supposedly detected by the birds was small, their stellar and inclination compasses would have to be quite accurate. One way to test whether the birds really can measure declination differences would be to see how they respond if the stars are hidden from them, or if the rotational center of the stars is moved in a planetarium. Ideally, the Rybachy experiments would be replicated on free-flying birds with GPS trackers, though this would be technically challenging.

Although the matter is not yet settled, we now have, for the very first time, strong if not decisive evidence that a bird can solve the longitude problem using geomagnetic and celestial cues in parallel.

How can salmon, fattening themselves up on the plentiful food supplies of the open ocean, locate the estuaries of the rivers in which they were born, especially when these may be several thousand miles away?

One of the virtues of the geomagnetic field is its omnipresence: Wherever you are—on the land, in the air, or even under the sea—you can, with the right sensors, detect it. Since salmon can orient in earth-strength magnetic fields,[8] the notion that the salmon's trans-oceanic homing system might depend on geomagnetism is appealing. But it is obviously not easy to conduct experiments on fish cruising around in the open sea.

Nathan Putman discovered that records of sockeye salmon catches had been kept over a fifty-six-year period to help settle disputes between the Canadian and US authorities about how they should be shared between the two countries. Of particular interest to him were the salmon that breed in the Fraser River of British Columbia. This enters the sea just north of downtown Vancouver, 850 miles from its source high in the Rocky Mountains.

These fish typically spend two years in the open ocean before returning to spawn. Confronted by the long palisade of Vancouver Island, they face the choice of approaching the mouth of the Fraser River either from the north, through the Queen Charlotte Strait, or from the south, via the Strait of Juan de Fuca.

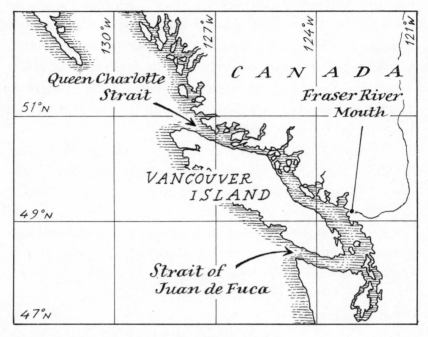

Salmon returning from the open waters of
the Pacific to spawn in the Fraser River
follow either of two different routes:
Queen Charlotte Strait or the Strait of
Juan de Fuca.

The fisheries' records revealed intriguing annual variations in the numbers of salmon arriving from each direction. That information by itself was of no help, but Putman also knew that the geomagnetic field around Vancouver Island was subject to gradual changes, known as "secular drift." He wondered whether a comparison of the two processes—the variation in catches and the secular drift—might shed light on how the fish were finding their way.

Putman found that the fish favored approaching the Fraser River by the passage at which the magnetic intensity differed least from that around the river mouth. It looked as if the fish imprinted on the magnetic signature of the river as they left it and, when homing toward it, employed some kind of magnetic intensity sensor to select the route they followed. Some years this meant that the fish used the southern Juan de Fuca route, while in other years they preferred the northern one through the Queen Charlotte Strait.

You may wonder how the salmon can make use of magnetic intensity gradients, given that the intensity signal is so noisy and imprecise. But salmon are not homing pigeons; they only need to choose between two wide channels separated by a few hundred miles, so great accuracy is not required. Putman thinks that the fish might be using a magnetic map when navigating homeward across the open sea from their feeding grounds in the Gulf of Alaska.[9]

But by the time the fish are nearing the mouth of the Fraser River, they may well be relying not so much on magnetic as on olfactory information. Putman has since carried out further experiments, which suggest to him that *juvenile* salmon may use a combination of magnetic intensity and inclination signals, to set course for their mid-ocean feeding grounds when they first put out to sea.[10]

Putman's findings are fascinating, but the evidence that salmon have access to magnetic maps is not conclusive. As in the case of the Russian birds, the possibility that the fish in these experiments were really using a simpler mechanism—perhaps based on magnetic landmarks or beacons—cannot yet be ruled out.

Startled deer tend to run off together in a group, all bounding along in the same direction. This probably means that they have a better chance of avoiding collisions and can more easily reassemble once they are out of danger. But how do they all decide which way to go?

In trying to answer this question, scientists have recently frightened 188 separate groups of roe deer in various hunting grounds around the Czech Republic.[11] What they found was that—even when account was taken of other likely factors, such as wind and sun direction—the deer preferred to seek refuge in a magnetic northerly or southerly direction. If the threat appeared from the south or north, they would go in exactly the opposite direction, while if it came from the east or west, their escape route would tend toward either north or south. They avoided escaping in an easterly or westerly direction if they could. It also emerged that, when grazing peacefully, the deer tended to be aligned along a magnetic north-south axis.

These findings suggest that roe deer are sensitive to geomagnetism and that they use it to coordinate their escape behavior—the first time this has been shown in any mammal.

Chapter 20

THE MYSTERY OF SEA TURTLE NAVIGATION

The sight of a female sea turtle laboriously dragging herself out of the sea and up the slope of a sandy beach to build her nest is almost unbearably touching. All that effort and devotion are a potent symbol of motherhood, or if that is too anthropomorphic for you, the overwhelming strength of every animal's reproductive urges.

But for animal navigation scientists, female turtles are fascinating for another reason: They are remarkably good at homing, and it now seems clear that they rely heavily on magnetic cues to find their way.

As well as being an expert on pigeons, Paolo Luschi is one of the small group of scientists who have conducted extensive research on turtles in the wild. This has usually involved attaching tracking devices to their shells when they come ashore to nest. When I saw Luschi in Pisa, he told me about the challenges involved in work of that kind.

Turtles are big, strong animals; green turtles, for example, are around a meter long and may weigh 450 pounds or more. When they emerge from the sea, which they usually do at night, they pull themselves up the beach using their front flippers, to the point where the vegetation begins.

Once they have found a suitable site to build their nest, they start by scooping out a shallow depression known as the "body pit." Then, with surprising dexterity, they construct a roughly cylindrical "egg chamber" ("a very nice piece of architecture" as Luschi describes it), using their rear flippers one at a time to remove the sand. Quite often, if the turtle is unhappy with the results, she will either give up and return to the sea or start all over again—very frustrating for the waiting scientists.

When they are satisfied with the egg chamber, the turtles start laying and typically deposit eighty to one hundred eggs, each roughly the size of a table tennis ball and soft to the touch. Once laying has begun, they do not stop and no longer show any fear. This is the purpose of their lives. In fact, it is almost impossible to distract them; at this point, Luschi says, "you can do anything with them."

This is the moment the researchers have been waiting for, but they have to move quickly, as the egg-laying may only last for half an hour. Before they can attach the tracker, they have to clean the shell, first with sandpaper, and then with acetone. Then they glue it to the shell with a water-resistant epoxy resin. The turtles do not seem to mind.

When the turtle has finished laying, she carefully covers her eggs with sand using her hind flippers. Then she rapidly fills in the body pit using her powerful front flippers. Sand now flies everywhere, and the researchers have to take care not to get whacked, which can be painful. The aim of the turtle is to conceal the nest site from potential nest raiders, and once it is fully covered, it heads straight back to the sea. If the resin has not had time to dry, it may be necessary to stop the turtles getting back in the water.

Doing so by force is far from easy, as they are very determined; it is a bit like trying to stop a "small tank," and it takes two or three people to block their progress. But there is no need for that. All you have to do is show the turtle a flashlight and they will follow it. It is, Luschi says, a bit like taking a large, slow dog for a walk.

Over the last thirty years or so, scientists have revealed navigational abilities in these marvelous reptiles that are at least as impressive as those of salmon, though until the 1950s, they were the subject of folklore rather than science.

There were plenty of fishermen's tales about turtles returning to the beaches where they were born, but very little was known for certain about how they lived—other than the fact that they nested periodically on certain beaches, and traveled far and wide in between. The main reason people were interested in them was that they were extremely good to eat. The Lord Mayor's Banquet—a lavish annual dinner held in London for the rich and powerful—always used to include turtle soup, which was a great delicacy. It has long been off the menu in London, but turtles and their eggs are an important source of income (and protein), and many people in the tropical countries where they commonly nest depend on turtle meat and eggs for their livelihoods. This can give rise to awkward tensions between conservation requirements and human needs.

One of the first scientists to study turtles in the wild was Archie Carr (1909–87). He was an influential conservationist, long before the protection of the natural world had become a popular cause, and he played an important part in persuading the authorities to set up a national park at Tortuguero, on the east coast of Costa Rica—the world's first turtle sanctuary. A wildlife refuge bearing his name has also been established on the east coast of Florida.

In order to find out more about what green turtles got up to when they left their nesting beaches, Carr first tried tracking the females by attaching balloons to them. This only worked over quite short distances and so, following the example of the birders, he started tagging them. This did not go well. The earliest tags were attached to the animals' shells by strong wire, but these often came off even before the turtles had left the nesting ground.

Although much was mysterious about what the female turtles were doing in the intervals between their recurrent bouts of egg-laying, one thing was certain: "a lot of strenuous romance" was going on just offshore, and it soon became clear that the loss of tags was the work of the rutting males:

> Sea turtles in love are appallingly industrious . . . To hold himself
> in the mating position on top of the smooth, curved, wet,

wave-tossed shell of the female, the male employs a three-point grappling rig, consisting of his long, thick, recurved horn-tipped tail, and a heavy, hooked claw on each front flipper. Sea turtles breathe air, of course . . . so both sexes naturally try to stay at the surface during the violent mating engagement. This adds to the acrobatic problems of the male, and augments his intemperate scraping and thrashing at the shell of his intended . . . During that time the other males gather, and all strive together over the female in a vast, frothy melee, in which nothing can be seen from the shore except that it is pretty exciting.[1]

Eventually, Carr started using cattle tags attached to the front flipper, rather than the shell, and these worked much better. But he attributed much of the success of his tagging program to the reward of $5 offered for each returned tag. This was quite a lot of money for a Caribbean fisherman in the 1950s and, crucially, more than the value of a turtle at market.

As more and more tags came in, Carr was able to prove that the apparently tall tales of turtle migration and homing were fully justified. How the turtles found their way when out at sea was a great puzzle that he was unable to solve, but he made an important first step by defining many of the key questions that needed to be answered.

Of particular interest to Carr were the green turtles that travel from their feeding grounds on the coast of Brazil to breed and lay their eggs on the beaches of Ascension Island. A "crumb of land out in the sea between Africa and South America," Ascension is so small and remote, sometimes even human navigators have had trouble finding it. During the Second World War, aircraft being delivered from the US to Burma, via Brazil and Africa, used it as a refueling stop—but if they failed to locate the island, they would be forced to ditch in the South Atlantic. There was a saying among the pilots: "You miss Ascension and your wife gets a pension."[2] No doubt that concentrated the minds of the navigators.

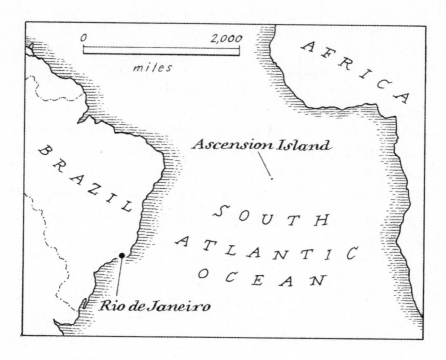

Ascension Island's remote location

How then do the turtles find Ascension? Carr realized that visual landmarks would be irrelevant for most of the 1,400-mile journey, though the volcanic peak (altitude 859 meters) at the center of the island might perhaps be visible to the turtles from quite a long distance. He was well aware of the discovery of a compass sense in insects, and in other animals, and wondered whether the female turtles might have a similar faculty. But what chance would there be of hitting "a five-mile-wide target after a thousand-mile swim, if she guided herself by compass sense alone?"

Even if there were no currents to contend with, Carr thought this would be an incredible feat of wayfinding. Since the turtle will be facing a steady west-going current, not to mention occasionally rough seas, Carr concluded that a compass alone simply could not get the job done: "There has got to be something more to the navigation process than that. The great enigma that one day has got to be solved and explained is how animals . . . find all the islands that they regularly visit."[3]

Carr speculated about the possibility that some odor or taste emanating from Ascension Island might act as a kind of beacon, spreading out across the sea. But it seemed unlikely that this was the answer, as the turtle would be forced, like a moth, to follow a long, exhausting zigzag course as it tracked the odor plume to its source. He wondered also whether the turtles might be following contours on the sea floor, or maybe homing in on sounds—perhaps the extraordinarily loud noises produced by snapping shrimps—though he did not consider infrasound.

Other mechanisms like inertial or celestial navigation were possibilities, though there was no evidence to go on. He even considered the so-called Coriolis force. Perhaps by detecting the slight changes in acceleration caused by the rotation of the earth, as it moved north or south, the turtle could estimate its latitude. But this seemed too far-fetched. Finally, he discussed the possible role of geomagnetism. Though at the time (the mid-1960s) there was no hard evidence that any animals could navigate using magnetism, he rightly suspected that this was a promising line of inquiry.

Green turtle homing

It fell to the next generation of turtle students to address the fascinating questions that Carr had raised, and among them, Floriano Papi and his pupil, Paolo Luschi, were to be prominent figures.

Although famous for his work on olfactory navigation in pigeons, Papi always insisted that he was not an ornithologist but an ethologist; it was how animals navigated that fascinated him, rather than the behavior of any particular animal. He was also excited by the new tracking technology that was starting to become available in the late 1980s.

Papi happened to meet two Malaysian researchers at a conference in the early 1990s. They were using short-range radio transmitters to track turtles and Papi—a man of restless curiosity—was so intrigued that he decided to start investigating turtle navigation himself. Luschi, who had only graduated in 1989, was working on pigeons at the time, and was taken completely by surprise when Papi mysteriously asked him whether he would like to make a trip to the tropics. Luschi could hardly say no to an offer like that, especially when Papi finally revealed what he had in mind.

So, in 1993, the young researcher headed off to the remote island of Redang, off the east coast of Malaysia, to do his first experiment on turtles. He had never traveled outside Europe and was entranced by the unspoiled beauty of the beach where the green turtles came ashore to nest—something they do repeatedly over a period of several months.

The timing of this first expedition was determined by the Italian summer holidays. July was not the best moment to choose, since it was the middle of the turtle mating season and there was a strong probability that any tracking device attached to a female would soon be ripped away by an overexcited male. The trick was to find a female at the very end of her egg-laying cycle, who would, on leaving the beach, head straight out to sea.

To complicate matters further, the first tracking devices they tried turned out to be hopelessly leaky and broke down. But, as Luschi says, Papi was a lucky man. Despite all these difficulties, the team managed—with the help of their Malaysian colleagues—to obtain some of the earliest satellite tracking data from a migrating green turtle.

One particular female traveled more than 370 miles from the nesting beach to her feeding grounds far out in the South China Sea. Even more impressive than the distance she covered was the fact that she maintained a steady course for the last 300 miles of her voyage.[4]

Obtaining accurate positional information from an animal that spends most of its time underwater is tricky. The transmitters that Luschi has usually employed need a few seconds to transmit enough data to the satellites, and they can only do this when the turtle surfaces to breathe, which it does only briefly. So fixes may be few and far between and, even in the best circumstances, are not very precise.

But the tracking devices can now be coupled with GPS to provide a much more accurate position.

Coming from a background in pigeon studies, Papi was naturally keen to attempt displacement experiments on turtles. In 1994, Luschi, still only a PhD student, went back to Malaysia, but this time without Papi. He and his team succeeded in tracking a displaced female green turtle returning to her nesting beach and later plotted the routes followed by several turtles on much longer migratory journeys.

The results were astonishing: One of these turtles swam from Malaysia all the way to northern Borneo, and another to the southern Philippines. Once again, they found that the turtles followed courses of arrow-like straightness—this time over distances that easily exceeded 600 miles.[5]

Papi and Luschi next went to South Africa, where they worked both on loggerhead turtles and the huge leatherbacks—magnificent creatures with deeply ridged, leathery backs that are about the same size as an old Fiat 500 car. This time they collaborated with George Hughes, the head of the Natal Parks Board, who had been tagging turtles since the early 1960s.

In one displacement study,[6] they showed that female loggerheads could find their way back to their nesting grounds over distances of up to 43 miles. Later they tracked a single leatherback on a journey of more than 4,000 miles, which included a long period during which the animal traveled in an almost straight line, though that may have been partly due to the effects of a strong ocean current.[7]

Luschi and his colleagues started exploring the navigational abilities of freely moving green turtles, when they later went to Ascension Island. As is so often the case in field work, the results were not clear-cut. In one displacement study,[8] they captured eighteen females on the island, attached trackers, and released them in the open sea at distances ranging from 37 to 280 miles—not very far by turtle standards. Four of them headed straight for Brazil (where their feeding grounds lie), four others eventually headed in that direction, but only after circling around for a while, and only ten went back to Ascension.

The navigational performance of those returning to Ascension was poor. All but one of the recorded tracks were circuitous, though the final segments were straight, "as if the turtles were searching for a sensory contact with the island which they obtained at various distances." Most approached the island from a downwind direction, which led Luschi to suggest that they were relying on some "wind-borne information" carried from the island—possibly an odor plume.

A later study pointed more clearly toward the importance of olfaction in green turtle homing.[9] On this occasion, female turtles were taken from a nesting beach on Ascension Island, fitted with satellite trackers, and transported by ship to locations that were thirty miles away—in either an up- or a downwind direction. The turtles released downwind all made their way back to the island within a few days, while those taken upwind had far more trouble relocating it.

In fact, one of the upwind turtles had still failed to locate Ascension after fifty-nine days of tracking, despite the fact that it had earlier approached within sixteen miles of the island. It seems quite likely that an odor plume blowing from the island helped the downwind turtles find their way back to it, though the evidence is not decisive.

Luschi later carried out a challenging study in the Comoro Islands, a remote archipelago in the Indian Ocean, between Madagascar and Africa. His aim was to find out whether an artificial magnetic field affected the homing performance of green turtles.[10] The nesting beach could only be reached by sea, and Luschi sailed there in a small yacht, which was not much fun for a man who suffers badly from seasickness. The team managed to get the turtles on board the yacht using a makeshift litter, though not without difficulty; it helped that one of the team was a burly rugby player. But then strong winds prevented them leaving the shelter of the lagoon.

Luschi was already feeling unwell while they waited for the winds to ease, but things got much worse when they headed out to sea. It took twelve hours to reach the release site, and he was severely seasick all the way; by the time they got there, he was barely able to stand. On the return journey to Mayotte, running low on fuel, they had to resort to the use of sails. Though it was a relief to be spared the noise of the engine, the journey lasted much longer than planned, and since they had no radio-telephone, they could not warn their colleagues of the delay. Luschi and his crew were delighted to get back on dry land and the shore-based team was greatly relieved to see them.

On this expedition, Luschi and his French colleagues released twenty turtles at sites sixty to seventy-five miles away out in the Mozambique Channel. Thirteen had magnets attached to their heads. All but one managed in the end to get back to Mayotte, though not always by a direct route; they seemed unable to make allowances for the effects of ocean currents. But the turtles encumbered with magnets followed much longer homeward paths. This was the first evidence from the field that turtles might actually make use of magnetic cues for navigational purposes.

∞

Many marine fish lay eggs, which hatch to produce larvae that float around freely as plankton. The likelihood that these might eventually grow into adult fish able to find their way back to where they were conceived might seem rather low. Yet that is precisely what seems to happen with Atlantic cod.

DNA fingerprinting techniques applied to the bones inside the ears of fish (otoliths) now enable scientists to determine exactly where the animals started life. A recent analysis looked at archived otoliths collected (over a period of sixty-odd years) from cod tagged in the waters off the west coast of Greenland. It revealed that 95 percent of the tagged fish that had later been recaptured off Iceland originated in Icelandic waters.[11] They therefore must have traveled to Greenland (where they were tagged), before returning to Iceland. This is convincing evidence that cod can home successfully over distances of 600 miles or more.

Though we have no idea as yet how cod manage to find their way home, the fact that they do is of great importance to fisheries management.

Chapter 21

COSTA RICAN ADVENTURES

Ken Lohmann, a professor at the University of North Carolina, Chapel Hill, is soft-spoken and perhaps a little shy. Answering a question, he often begins rather hesitantly, "Hmm, now let me see," as he gathers all the facts. But nobody knows more about how marine turtles make use of magnetism, and the extraordinary discoveries he has made over the last thirty years now form part of the bedrock of animal navigation studies.[1]

I was lucky enough to spend a whole week in Lohmann's company, when we joined two of his doctoral students—Roger Brothers and Vanessa Bézy—who were conducting experiments on the Pacific coast of Costa Rica. Roger and I traveled on the same flight from Miami and we met Vanessa at Liberia airport. She was carrying a large radio antenna, which we only just managed to squeeze into the jeep I had hired.

After a surreal, jet-lagged, open-air lunch at a German bakery not far from the airport, we headed south for Playa Guiones, a beach resort popular with surfers, some seventy-eight miles to the south. On the way, Vanessa and Roger told me about the work they were doing, the various problems they had been trying to overcome, and the questions they hoped to answer. Ken arrived separately a few days later.

Female olive ridley turtles periodically come ashore to nest— hundreds of thousands of them—on a long, grey beach beside the small village of Ostional, about six miles north of Playa Guiones.

In Spanish, these extraordinary events are known as *arribadas* (arrivals) and they were completely unknown to science until modern times. For the local people, however, they have long been a vital source of income, as they can sell the turtles' eggs for high prices. Nowadays there are strict controls on egg collection, but it is still permitted at certain times. The sand, which has a rich smell, is filled with fragments of turtle eggshell. Vultures and caracaras (large falcons) are on constant watch for hatchlings.

The female turtles coming ashore during an *arribada* are so numerous that they clamber over each other, in their search for a clear space in which to build their nests, and quite often one turtle will accidentally dig up the nest of another. If the aim were simply to overwhelm potential predators, they would not need to come ashore in such absurdly large numbers. Nobody yet understands why the *arribadas* take place and they seem to make very little sense, but they certainly demonstrate the remarkable homing ability of these turtles.

They generally last for a few days at a time and are not rare events, occurring fairly regularly at Ostional (though hardly anywhere else in the world) during most months from June through to December. *Arribadas* usually coincide with the third quarter of the moon, a fact that raises interesting questions about how the turtles keep track of time.

Right by the beach at Ostional, in the shelter of a few bushes next to the park rangers' station, Roger had built from scratch—largely using materials obtained from local hardware stores—an outdoor magnetic coil system designed to generate a uniform field around a circular, plastic water-filled arena. It was here that he planned to explore the possible role of geomagnetism in the natal homing behavior of the olive ridley turtle, with the permission of the local park rangers.

The turtles, which are smaller than the green, would have to be captured and manhandled up to the arena, and Roger was counting on Vanessa, Ken, and me to help him. Vanessa, for her part, was trying to find out what triggered these extraordinary mass nesting events. Her plan involved attaching radio transmitters to female turtles while they were still out at sea, and then tracking them as they came ashore.

Working from a small open boat, she and Roger had already managed to tag a number of turtles, but the first radio antenna had proved useless. She hoped the new one would work better.

It was the tail end of the Central American wet season. The journey from the airport was easy at first, but for the last thirty miles or so, the roads were full of huge, water-filled potholes that had to be negotiated with care, even in a four-wheel-drive jeep. The swirling, milk-chocolate rivers were close to bursting, and heavy surf broke on the long sandy beaches. The sea itself was stained brown and filled with floating tree trunks and other debris. Even Vanessa, who lives in Playa Guiones, had rarely seen such bad conditions. Both she and Roger were gloomy, and when at last I got to bed, exhausted after my long journey from London, I was feeling dispirited, too.

It was still dark when the lost-soul cries of the howler monkeys in the tall trees outside my window woke me abruptly on the first morning. The rain was still falling and when we tried to reach Ostional, we failed even to get across the first river, but within a couple of days, the clouds lifted. Steam now rose from the ground under the hot tropical sun, huge iguanas lumbered out of their hiding places, and gorgeous butterflies, including the occasional iridescent-blue *Morpho*, flapped among the flowers. Now at last we were able to ford all the rivers and bounced our way along the dirt road to Ostional.

Vanessa was using a programmable drone with a video camera to monitor the turtles, as they lolled in the waves a few miles out to sea, mysteriously biding their time. This marvelous device would follow a precisely determined course, and then return obediently to the point where we were standing, gently landing beside Vanessa like a trained falcon. It was easy to see the female turtles on the live video feed—lots of them, but day after day went by and still nothing happened. It was frustrating, especially so for Roger and Vanessa, and a telling reminder of how uncertain life can be for the scientist working in the field.

Although the odd turtle came ashore to nest and we often saw the tracks of the hatchlings heading down toward the sea, I eventually had to return home without seeing the big event. Though disappointing, the non-arrival of the turtle armada was in a way providential. Rather than working flat out every night wrangling turtles, and catching up on sleep during the day, we all had plenty of time to talk.

Ken Lohmann grew up in Indiana—about as far from the sea as you can get in the US—and, as a young boy, he was fascinated by the monarch butterflies that flew by his home in large numbers. But family holidays took him to the ocean, and there he was entranced by the strange creatures he discovered in the tide pools. His interest in marine animals grew while he was an undergraduate studying biology at Duke University.

After completing his master's degree in Florida (where he explored the subject of magnetic navigation in spiny lobsters, a subject to which we shall return), Lohmann moved to the opposite corner of the country: a marine laboratory in the glorious setting of the San Juan Islands in the Pacific Northwest. He was still fascinated by the elusive magnetic sense, but now he had to settle for a subject that thrived in the cold northern waters: a big, pink sea slug called *Tritonia*. This apparently unpromising animal offered one great advantage: It could easily be studied in a laboratory setting.

Lohmann started to make electrical recordings from individual cells within the slug's nervous system, and he made the surprising and important discovery that it was sensitive to changes in the surrounding magnetic field. In fact, it seemed to have a magnetic-compass sense. Having obtained his doctorate, Lohmann now began working on turtle navigation under the inspiring leadership of Mike Salmon, an expert field biologist.

When hatchling turtles emerge under cover of darkness from their nests in the sand, the first challenge they face is getting safely into the sea. Raccoons, crabs, and foxes like nothing better than to make a feast of baby turtles, so it is vitally important that they find the shortest route to the water's edge.

When they first emerge from the nest, the tiny turtles scuttle over the sand like little clockwork toys in an effort to reach the sea before they are eaten. They rely mainly on visual cues to get to the water's edge—being drawn toward lights that are low in the sky—so it is easy to see why the bright lights associated with a human presence cause such havoc among newborn turtles. The hatchlings also prefer to go downhill, which makes sense, since beaches slope toward the sea.

If the hatchlings succeed in reaching the water's edge, they immediately start swimming like crazy and continue doing so for a day or two, fueled by a small amount of yolk left over from the egg. After bashing through the breaking waves, they need to get offshore as quickly as possible, to escape the many marine predators that lie in wait for them in the shallows. Once they are well clear of the coast, they are swept up in the north-going Gulf Stream and embark on a roughly 9,000-mile journey that will take them around the whole North Atlantic basin.

Finally, perhaps after several years of oceanic wandering, they will return as juveniles to feeding grounds near the beaches where they hatched, and in due course, the females among them will mate and lay their eggs on those very beaches.

The first question that Lohmann and his colleagues addressed was: How do the turtle hatchlings get clear of the beach? With what he describes as the "typical hubris" of a young post-doc, Lohmann thought it was "obvious" that the hatchling turtles must be setting their course with a magnetic compass. After all, if a sea slug had one, why not a turtle? This was in 1988, and it turned out to be the start of a long, fascinating story, the end of which has yet to be reached.

Salmon devised a "floating orientation arena," which enabled them to test which way the hatchlings preferred to go once they entered the sea. The researchers would go out in a boat, maybe twelve miles or more, and drop the arena into the water. The hatchlings could not see the land from that distance, but they always seemed to keep heading eastward, toward the open sea.

This encouraged Lohmann and his wife, Catherine (a fellow scientist and frequent collaborator), to believe that they were indeed using a compass, but then—by happy chance—they encountered a few days of glassy calms. Now the turtles began swimming in circles, as if totally disoriented. When the wind picked up, the turtles started heading east once more. This was puzzling; perhaps, after all, magnetism was not the key.

It was starting to look as if the heading chosen by the hatchlings was actually determined by the direction in which the waves were traveling. This was confirmed by experiments in wave tanks, but there remained the possibility that they might be following a gradient, perhaps based on some kind of scent that was being blown toward the land. To eliminate that hypothesis, Lohmann needed a day when the wind was not blowing onshore. Then the hatchlings would be forced to choose between following their normal offshore course or responding to the waves.

The passage of Hurricane Hugo in 1989 gave the scientists the opportunity they needed. One morning, they woke to find that the wind was blowing strongly from the west—from the land toward the sea. The Lohmanns rushed out with their hatchlings and let them loose in the choppy waters off the east Florida coast. Sure enough, in these conditions the little critters headed for the shore.[2] This was the clincher: Wave direction was indeed the key.

The hatchlings might perhaps *look* at the waves to work out which way to go, but since they normally enter the sea in darkness and swim under water, only surfacing occasionally to breathe, that would not be at all easy. In fact, the explanation is a good deal more complicated than that. Lohmann eventually discovered that they are sensitive to the characteristic rotational accelerations—upward, backward, downward, then forward—to which they are exposed *within* an approaching wave. This they demonstrated by harnessing the hatchlings in a "goofy-looking device" that reproduced these movements. It is a completely automatic response, which they even display by "swimming" in the air,[3] and subsequent experiments have shown that most—if not all—other turtle species behave in exactly the same way.

Though it was now clear that the hatchlings had no need for a magnetic compass at this very early stage in their lives, Lohmann and his colleagues remained convinced that magnetism must play an important part in turtle navigation.

His next challenge therefore was to establish whether hatchling loggerhead turtles showed any response to altered magnetic fields, when temporarily confined in artificial arenas—initially improvised from old satellite dishes and children's paddling pools. But before they could do any experiments, they had to devise a special harness that allowed the hatchlings to swim freely while suspended from a bar that extended over the arena, as well as a simple electronic system for tracking where they headed.

It was, Lohmann admits, "very, very tedious work," and one of the major problems they soon confronted was that in complete darkness, the hatchlings refused to swim consistently in *any* direction. Since the normal wind-driven waves of the sea were absent, this did not come as a big surprise, but they found that the hatchlings were "exquisitely sensitive" to any differences in light intensity. In fact, their tendency to orient toward any light source was so strong that it overwhelmed every other response.

Lohmann was confronted with a really big problem: If he worked in darkness, the animals would go in all directions, but if he showed them any light at all, they would doggedly head toward it and take no notice of any other cue. How then was he going to measure the effects of an altered magnetic field? He had to find a way of working around it.

Like the humpback whale, the northern elephant seal is another prodigious transoceanic voyager.[4] These huge animals shuttle back and forth each year between their rookeries on the Channel Islands off the coast of California and—in the case of the females—the Aleutian Islands. The males for some reason prefer their own company and head off separately to the Gulf of Alaska. The females cover a distance of at least 11,000 miles in the course of a year,

while the males go at least 13,000 miles, and the courses they all follow across the open ocean are astonishingly straight. Their navigational methods are just as puzzling as those of whales.

But it is not just large marine mammals that engage in long-range migration. Great white sharks have been tracked traveling right across the Southern Ocean from South Africa to Australia—and then back again.[5] Some members of the shark family are sensitive to magnetic fields,[6] so the possibility that they rely—at least in part—on magnetic information to navigate over large distances deserves to be taken seriously. But they are also extremely sensitive to olfactory signals, so these too could well be involved.

A recent analysis[7] of tracking data from humpback whales, elephant seals, and great white sharks suggests that gravity might even play a part in their navigational systems. The strength of the gravitational force varies across the earth's surface—especially in a north-south direction. The weight of an animal, and therefore its buoyancy, will change from place to place. A typical humpback whale will apparently require about 200 pounds less buoyancy to remain effortlessly afloat in a tropical habitat than in a high-latitude one. If the animals can detect such differences, then they could in theory derive useful navigational information from them—though allowances might have to be made for changes in the salinity of the water, since this would also affect the buoyancy of an animal.

Chapter 22

A LIGHT IN THE DARKNESS

Ken Lohmann's answer to his problem with the hatchling logger-head turtles was to turn their light-seeking behavior to his advantage. He kept the arena in complete darkness and then presented the swimming hatchling with a light in the east. Once it was reliably swimming toward it, he turned the light off and, without altering the natural magnetic field, monitored the results.

The young turtles doggedly continued swimming eastward, but when Lohmann reversed the direction of the field, the hatchlings turned around and headed west. His not unreasonable inference was that this 180-degree change in the animal's orientation must be due to the altered magnetic field, and if he was right about that, then it followed the loggerhead turtle must indeed have a magnetic compass. With some variations, Lohmann and his colleagues have continued to use this procedure in their hatchling research ever since.[1]

The azimuth of the light does not seem to matter. Apparently, in the early days Lohmann and his colleagues tried using a light in the west and found that the animals continued swimming westward after it was turned off—not the normal direction for a hatchling heading out to sea from the east coast of Florida. When the light was turned off, a reversed magnetic field made them turn around and head in the opposite direction, as was the case when they showed a light in the east.

But it is not immediately obvious how these experiments relate to the natural behavior of the hatchlings. When I asked Lohmann about this, he offered the following explanation.

When the hatchlings emerge from the nest, they follow the light, and with any luck this leads them to the water's edge. Having entered the sea, they orient at right angles to the oncoming waves, which are always parallel with the shore; but when they get into deep water, the waves no longer provide a reliable guide, because their direction is then determined mainly by the wind. At this point, the hatchlings switch over to their magnetic compass in order to maintain an offshore heading: "It may just be that the experience of holding a course—no matter what cue they use—is enough to allow them to transfer it to their magnetic compass."

Using a magnetic coil system, it is possible to adjust magnetic field intensity and inclination separately. Lohmann next began to explore how the hatchling's compass actually worked, and in particular what the respective roles of intensity and inclination might be. He first repeated the same orientation experiment, but this time, after the light in the east was turned off, he tried altering only the inclination of the field—making it three degrees steeper than it was at the home beach.

He expected that the animals would either head east as usual, or would be totally confused, and therefore randomly oriented. In fact, the turtles turned resolutely toward the south.[2] This was really puzzling:

> We were pulling our hair out for a while, trying to figure out what was wrong with our setup. We assumed there was a light leak or something else wrong. We tried over and over again to get rid of this bias.

But one night, Lohmann and his team had a careful look at a magnetic map of Florida and noticed something significant. They saw that the altered field to which they were exposing the hatchlings actually matched one that occurred naturally a bit farther up the coast to the north. Suddenly a light went on:

Wow! Maybe there's nothing wrong with the experiment. Maybe they're actually using the inclination angle as an indication of latitude . . . Up until that point we had not even been thinking about any part of the migration other than getting offshore to the Gulf Stream. And the dogma at that point was just that the turtles swam to the Gulf Stream, got in and then drifted passively with the current. No one was even too sure at that point that they came back to their home areas.

Lohmann has since shown that older, juvenile turtles virtually displaced to the north of their home feeding area also react by heading south, while those displaced to the south head north. This suggests that they, like the hatchlings, can also use magnetic inclination as a proxy for latitude.[3]

Once the hatchling loggerheads are caught in the powerful northerly flow of the Gulf Stream, they are carried out into the deep ocean and, over a period of years, during which—with any luck—they grow and prosper, they paddle along in the currents that form what is known as the North Atlantic Gyre. Provided they stay within it, this huge body of water, which circulates around the entire ocean basin in a clockwise direction, will eventually bring the now juvenile turtles back to within striking distance of their feeding grounds on the coast of Florida.

But unless they actively swim in roughly the right direction, they run a serious risk of straying outside the gyre—with probably lethal consequences. Computer modeling of the way in which "virtual particles" move within the gyre when driven only by the currents, as well as a comparison of the tracks followed by buoys and real live turtles, shows that young turtles must be anything but passive drifters.[4] But how do they know which way they should swim to stay in the gyre?

Having discovered that the hatchlings could use inclination to determine their north-south displacement, Lohmann began to explore what effect changes of intensity might have on their behavior. This time the results were even more surprising. When presented with intensity signatures like those to be found off the coast of North Carolina, the hatchlings generally headed east, but when the signatures matched those on the other side of the Atlantic—off the coast of Portugal—they headed west. In other words, at these two locations the young turtles seemed able to use intensity alone to head in a direction that would keep them safely on the conveyor belt of the gyre.[5]

Lohmann next altered both inclination and intensity in order to imitate the magnetic conditions that the hatchlings would encounter at various different stages on their journey around the ocean basin. So long as the hatchlings were "sent" to locations close to the edges of the gyre, they generally set off in a direction that would increase their chances of survival, and the headings they chose varied widely depending on where the virtual displacement placed them.[6]

So if they were sent to a point off the coast of Portugal, they showed a tendency to head south, while in the southern part of the gyre, they went in a generally northwesterly direction.[7] The data were quite "noisy"—in other words, the animals did not all obediently adopt exactly the same heading; that would be too much to expect, and in a more recent experiment they showed significant orientations only in certain parts of the gyre,[8] but the general point still stands.[9]

Nathan Putman (one of Lohmann's former students, whose work with salmon we have already discussed) has shown that hatchlings may be able to distinguish two widely separated locations that differ only in their longitude.[10] He displaced the baby turtles virtually to areas in the ocean either near Puerto Rico (longitude 65.5 degrees west, latitude 20 degrees north), or near the Cape Verde Islands (longitude 30.5 degrees west, latitude 20 degrees north).

When sent to Puerto Rico, they tended to head northeast, but when sent to the Cape Verde Islands, they headed southwest. Once again, these responses would help the turtles remain within the gyre. It is unlikely that the turtles are in this case relying on a single parameter, whether inclination or intensity, because neither of these varies much as you move from east to west across the Atlantic basin, though they do vary a lot from north to south. But the hatchlings could distinguish the Cape Verde and Puerto Rico sites if they were paying attention *both* to intensity and inclination.

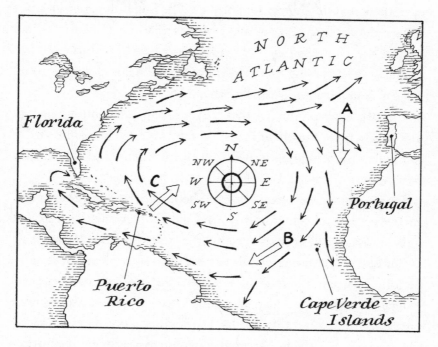

The North Atlantic Gyre. Turtle hatchlings in Florida "virtually" transported to different points around the Atlantic Basin (A, B, and C) responded by swimming in directions (marked with broad arrows) that would help to keep them safely within the gyre.

Lohmann and his colleagues interpret these findings as evidence that the hatchlings are born with an inbuilt sensitivity to the characteristic signatures of the earth's magnetic field around the gyre, defined by specific combinations of magnetic intensity and inclination. These signatures act like "open sea navigational markers" that trigger a hard-wired automatic response, which sends the turtles off in a direction that will tend to keep them well within the gyre. As in the case of Putman's Fraser River salmon, this system does not call for great precision; it is enough if the turtles can detect roughly where they are.

To judge from some of the wilder headlines that Lohmann's work has prompted, you might think it was an established fact that the turtles have their own biological version of GPS. But Lohmann does not believe the hatchlings have "any real idea of where they are." As he put it to me, choosing his words with typical care: "The turtles clearly can distinguish among the different magnetic fields along the route and can respond appropriately to them."

This formulation implies that they are map and compass navigators only in quite a restricted sense, but the idea that hatchling turtles can make use of magnetic fields, even in this limited way, is still astounding.

How could such a system be established? That is not a question that anyone can answer confidently. Turtles and their kin have been around for a hundred million years or more; they once breathed the same air as dinosaurs. Natural selection has therefore had plenty of time to perform its magic and must have favored the survival of animals carrying genes that enable them to identify key decision points along their migratory route. And the fact that they do not all respond in exactly the same way actually makes good evolutionary sense. The eccentricities that some animals display might enable the species to survive when the earth's magnetic field undergoes major changes, as, for example, in the case of a field reversal (*see page 124*).

Genetic techniques have confirmed that female turtles do indeed return to the area (if not the precise location) where their lives began to lay their eggs. A magnetic navigational system might explain how they manage this feat, and there is some evidence that the magnetic characteristics of the natal beach are a key factor in this process.[11]

Roger Brothers is pursuing the theory that the hatchlings—either while in the egg or just after hatching—"imprint" on the unique geomagnetic signature of the area around the nest, then find their way back to the natal beach years later, using this remembered information.

Following the example of Putman's work with salmon, Brothers has analyzed records of the locations of loggerhead nests in Florida over a nineteen-year period. Here again, as in British Columbia, secular drift means that the magnetic signatures of a given site (defined in terms both of inclination and intensity) gradually move along the coast.

If the imprinting hypothesis were correct, each turtle would come back to a location that differed slightly from the place where it was born. This in turn would result in predictable changes in the overall distribution of nests. Brothers therefore compared nesting densities at intervals of every two years (the typical gap between each female's bouts of egg-laying), while adjusting for fluctuations in the overall numbers of nests.

He found that nesting densities significantly increased in areas where secular drift brought magnetic signatures closer together, and decreased where it caused them to diverge.[12] Brothers's ingenious use of the nesting records lends weight to the theory that the homing behavior of the turtles is based on magnetic imprinting.

More recently, Brothers and Lohmann have shown that variations in the geomagnetic field are associated with genetic differences among populations of turtles nesting on separate beaches. Here then is the first genetic evidence of the reality of geomagnetic imprinting, and of its power to shape the structure of turtle populations.[13]

None of the experiments we have been discussing provides *direct* evidence that turtles in the wild rely on geomagnetic cues to find their way around. If you wanted to establish beyond question that young turtles circling the North Atlantic Gyre were responsive to geomagnetic "signposts," you would need to find a way of altering the field around them as they swam along in mid-ocean. Equally, to know for sure whether a female turtle was imprinting on its place of origin, you would need to alter the magnetic field around it at birth and then keep track of it—perhaps for fifteen years or more—to see where it eventually chose to lay its eggs.

If it then returned to the location defined by the artificial field to which it had been exposed at birth, you would have solid evidence that imprinting had taken place. Lohmann and his colleagues would love to be able to carry out experiments like these, but the problems they pose are simply overwhelming.

Although it is now very clear that magnetism plays a key role in the navigational behavior of turtles, and olfaction may well also be involved,[14] it is very likely that they also make use of other cues.[15] Perhaps—like Pacific islanders—they can use persistent ocean swells to maintain a steady heading. Maybe they can detect the characteristic wave patterns generated around oceanic islands, or home in on them by picking up characteristic odors, or listening for the sound of breaking waves. These are questions to which we do not yet have answers.[16]

Luschi likens turtle navigation to a process of *bricolage*—the French word for making something from any bits and pieces that come to hand. He thinks that turtles opportunistically make the best use they can of whatever useful information is available to them. They may even be able to judge which among the various sources of information available to them at any given time are likely to provide the most reliable information. But one thing does now seem clear: Even if they do not have access to magnetic maps, turtles do rely heavily on magnetic cues to guide them.

Amazing Crustaceans

Spiny lobsters (also known variously as langoustes, crayfish, or rock lobsters) are so different from us that they might just as well have come from another planet. Like turtles, they have been around for a very long time; a fossil ancestor has been found that is 110 million years old. They have ten long spidery legs and two very long antennae sprout from their heads. Most of us would probably be quite unaware of their submarine existence if they were not so good to eat. This unlucky characteristic has brought spiny lobsters (like their claw-fisted cousins) to the attention of fishermen, who catch them in large numbers. Strangely enough, they turn out to be some of the best navigators in the animal kingdom.

A nocturnal forager, the spiny lobster travels quite long distances on the hunt for clams and sea urchins before returning to the safety of its underwater lair. It also undertakes bizarre annual migrations, which take it from the shallows into deeper water to avoid the dangers of winter storms and hurricanes. The lobsters then form long conga lines, trudging along nose-to-tail for as much as 125 miles in a straight line—by day and by night. From our limited human perspective, spiny lobsters may not appear to be very talented animals, yet somehow they maintain a constant heading, despite the unevenness of the sea floor and the frequently poor visibility. That is an impressive navigational feat.

While studying for his master's degree in Florida, Ken Lohmann heard a lecture about the monarch butterfly's migration and the possibility that it used magnetic cues to set its course. Inspired by this, he spent some time trying to establish whether the navigational abilities of the spiny lobster might also owe something to magnetism. He was one of the first scientists to employ an electromagnetic coil to test whether an animal's behavior could be influenced by altering the magnetic field around it but, like most young researchers, he ran into problems. The first coil system Lohmann built burst into flames when he overloaded the circuits and, even when he managed to make it work safely, it proved difficult to generate a reliably uniform magnetic field around the captive lobster—essential if consistent findings were to be obtained.

Lohman's pursuit of the secrets of lobster navigation eventually led him to the SQUID—otherwise known as a Superconducting Quantum Interference Device. With the help of electrical circuits that are cooled almost to absolute zero, these machines are used to detect extremely weak magnetic fields. Lohmann now began chopping up spiny lobsters and lowering the chunks into a chamber surrounded by a vat of liquid helium to see if he could find any magnetically active tissue in them— and he succeeded in doing so. This was a thrilling result, but it was as far as he got; having obtained his master's degree, he headed off to study the sea slug, *Tritonia*.

Lohmann did not forget the spiny lobsters. Years later, he resumed his exploration of their navigational abilities, this time by conducting simple displacement studies. He and his colleague, Larry Boles, captured lobsters in the Florida Keys and transported them by boat to release sites up to twenty-three miles away. During these journeys, the animals were placed in opaque plastic containers filled with the same water they had been living in, so as to avoid giving them any telltale olfactory clues, and in case they were good at DR, the boat was driven around in circles as well.

Before releasing the lobsters, Lohmann covered their eyes with plastic caps and then tethered them in an arena where they could record the direction in which they moved. What he found was astonishing. The lobsters—far from being totally dazed and confused, as we would surely be in similar circumstances—reliably trundled off in a homeward direction. Assuming they had been unable to pick up any useful information *en route*, and could not detect any landmarks or beacons at the point of release, this suggested that they had some means both of fixing their position and of determining the correct homeward heading. This would count as "map and compass navigation": the Holy Grail of animal navigation studies.

Lohmann had already shown that spiny lobsters had a magnetic-compass sense.[17] So it was clearly possible that the animals in this latest experiment were keeping track of their movements on the outward journey using magnetic information. He therefore repeated the experiment—this time with a few additional twists.

The lobsters were now taken to the testing site by truck, and on half the trips the container they were in was lined with magnets, some of which were hanging from strings, so that they were constantly moving around. By disrupting the natural field around the lobsters, these would have denied them the opportunity to track their outward progress by magnetic means. On the other trips, they were transported in the same container, but without magnets. In every case the container was suspended by ropes, so that it swung erratically as the lorry was driven through a confusing series of turns and circles on its way to the test site.

Once again, when tested, the lobsters faithfully headed for home—regardless of whether or not they were traveling in the company of man-made magnets.

The next step was to conduct a virtual displacement with the same kind of magnetic coil system Lohmann had earlier employed on turtles. The physical displacements had been over quite short distances. Now they "sent" the lobsters on much longer virtual journeys: 250 miles, either to the north or south of their homes. Just like the juvenile turtles, they responded by heading roughly south or north, as if they knew which way they needed to go.

These extraordinary results suggested not only that spiny lobsters could perform map and compass navigation, but also that geomagnetic cues were at the heart of the process. How exactly the system works is less clear, though a combination of intensity and inclination may be involved. As Boles and Lohmann succinctly put it back in 2003 in their groundbreaking article in *Nature*:[18] "These results provide the most direct evidence yet that animals possess and use magnetic maps." That statement remains true today.

The salmon, the turtle, and the lobster are an ill-assorted trio—one fish, one reptile, one arthropod—but their very diversity is revealing. If representatives of such widely different animal groups all share an ability to make use of the earth's magnetic field to perform complex feats of navigation, it would be surprising if that gift were not more widespread. Whether the various different forms of magnetic navigation emerged at some very early stage in the evolution of life and proved so valuable that they have been widely conserved, or whether they have been repeatedly "reinvented," is as yet unknown.

Joe Kirschvink, a geophysicist at the California Institute of Technology, has recently caused a stir by reviving the previously discredited idea that humans, too, have a magnetic sense.

This theory was touted by a British scientist, Robin Baker, who claimed in 1980 that blindfolded students driven in a minibus along circuitous routes in the countryside around Manchester could (fairly) reliably indicate the correct homeward direction when they got off. In a follow-up experiment, the students carried either a small bar magnet or a similar-sized, nonmagnetic brass bar inside their blindfolds. Now the only ones who could correctly point their way home were the ones carrying the brass bars.[19] Baker took this as a strong indication that the directional sense was based on magnetic information and, not surprisingly, his claims attracted a lot of publicity.

But repeated efforts to replicate Baker's results proved unsuccessful, and a consensus emerged that the students in the original studies must have had access to some nonmagnetic directional information. In one particularly rigorous test, 103 Australian university students were outfitted in surgical overalls, mittens, and face masks, while their ears were muffled, perfume was applied beneath their nostrils and—final indignity—a blackout basket was put over their heads. These unfortunates pointed in random directions at the end of their journeys, though when they repeated them without all the encumbrances, they were able to indicate which way was north.[20]

Though Kirschvink was among those who called into question Baker's findings, he himself has recently claimed—on the basis of recordings of electrical activity in the brain—that humans can detect a change in the orientation of a magnetic field, even though they have no conscious awareness of it.

I was present at the conference (in 2016) at which Kirschvink first announced these findings and can vouch for the fact that they were greeted with some skepticism, though no one doubts his expertise and scientific credentials. His findings have now been formally published,[21] and if he turns out to be right, we will be left with a new puzzle. Perhaps this inaccessible sense is merely a useless vestige of a tool that was employed by our distant ancestors, but a magnetic compass would surely have been of inestimable value in hunter-gatherer societies, so why would natural selection have failed to preserve it?

One of the most mysterious of all migratory animals is the European eel. These extraordinary fish have a very complicated life cycle, which involves not one but two transoceanic migrations—however, their numbers have plummeted in recent years. In order to conserve them successfully, we need to understand their migratory behavior better.

Eels start their lives in the Sargasso Sea, a large area of ocean in the southwestern Atlantic. The first challenge facing the newly hatched eels (or leptocephali) is to enter the Gulf Stream, which will carry them—just like the hatchling loggerhead turtles—around the North Atlantic Gyre. When they reach the European continental shelf, where the water gets much shallower and less salty, they turn into glass eels and find their way into rivers and streams.

They then change into yellow eels, the adult form. As yellow eels, they may live up to twenty years before their sexual organs mature, triggering their return to the spawning grounds of the Sargasso Sea—roughly 3,000 miles away.

In a recent experiment,[22] glass eels were caught after they had entered the River Severn in Wales and subjected to a variety of magnetic displacements. These revealed that they were sensitive to "subtle differences in magnetic field intensity and inclination angle along their marine migration route." Moreover, it appeared that they were inclined to swim in directions that would increase their chances of entering the Gulf Stream—again, exactly like the hatchling turtles.

The weakness in this study is that glass eels are quite different from leptocephali. It is therefore uncertain whether these findings are relevant to the behavior of the newly hatched eels far out in the Atlantic.[23] Nevertheless, if eels are responsive to different configurations of the geomagnetic field at any point in their lives, it seems likely that they are indeed magnetic navigators. More research is plainly needed.

Chapter 23

THE GREAT
MAGNETIC MYSTERY

The hunt is on for the sensors that allow animals to detect the earth's magnetic field. Over the last decade or so, this challenge has brought together scientists from the fields of quantum physics and chemistry, geophysics, molecular and cell biology, electrophysiology, neuroanatomy, and, of course, behavioral studies, but the net may have to be cast wider still. A Nobel Prize may well reward those who eventually find the answers.

When scientists talk about visual, auditory, inertial, or olfactory navigation, they have a pretty good idea of the sensory mechanisms that are involved. They know what eyes, ears, and noses are like and how they work, though plainly the details vary enormously between the various animal groups. Shearwaters and dung beetles both use their eyes to see, but they see different things; a salmon can taste chemicals in the water that would mean nothing to a bird or a moth; and bats can do things with their ears that few other animals are known to do. In some species, scientists also have a good grasp of exactly how the nerve signals from the sensory organs are processed in the central nervous system—right down to the patterns of firing in individual brain cells.

But when it comes to geomagnetic navigation, the picture is much more confused. At present there are three radically different theories, any or all of which may prove to be correct. And it still cannot be ruled out that some entirely different mechanism, as yet undreamt of, may be at work.

This is such a complex and highly technical subject that I can only offer a brief summary of how matters stand.

One of the problems facing scientists interested in how animals sense the earth's magnetic field is that it easily penetrates living tissue. That means a magnetoreceptor need not be on the surface of the animal—like an eye, nose, or ear—it could just as well be buried deep inside it. Nor does it have to be large. It might not even be in one place: single cells might be at the heart of the system, and they could be scattered around the body—literally from head to tail. So there may, in fact, be no identifiable structure to find.

But the situation is not totally hopeless. We do know how magnetotactic bacteria respond to magnetic fields, and we also know that they have been around for a very long time. They carry within them microscopic crystalline chains of magnetite, which enable them to align themselves with the magnetic field around them in a completely passive way—like the needle of a compass. If the ability to detect the earth's magnetic field improves their chances of surviving and reproducing, it is possible that many or perhaps most animals have inherited a magnetite-based mechanism.[1] But how would this work in a multicellular organism?

An array of a few million cells containing magnetite could, it seems, be used to detect small changes in the intensity of the earth's magnetic field.[2] Reliable evidence of the presence of magnetite in an animal is hard to come by, because it is extremely easy to contaminate tissue samples—even airborne particles of volcanic dust can cause problems—but it has been found in insects, birds, fish, and even humans.

The ubiquity of magnetite suggests that it must be doing something important. Honeybees, for example, have magnetite-based permanent magnets in their abdomens. These start to form when the insects are still at the larval stage, and presumably acquire their orientation when each one is cosily confined within its own individual cell as a pupa, lined up at right angles to the surface of the honeycomb. Honeybees are also equipped with hundreds of specialized cells in their upper abdomen, which contain thousands of separate grains of magnetite. These cells are embedded in a matrix that is thought to expand or contract in response to changes in the surrounding field. Some believe that this mechanism may provide the bees with an inclination compass.[3]

Trout can quickly learn to obtain food by running their noses against an underwater target, which can only be identified by detecting a slight alteration in the intensity of the surrounding magnetic field. This ability apparently depends on magnetite contained in cells in the fish's nose—and similar cells are also found in salmon. (The trout show no sensitivity to changes in field inclination.) And sharks that learn to detect and approach magnetic targets seem to be relying not on their well-known electrical sensitivity but on a separate magnetic organ.[4]

In 2007, it was announced that sensory nerve-cell endings in the beaks of pigeons contained magnetite and another magnetic material.[5] Since the only nerve serving this part of the pigeon's anatomy is the trigeminal, it was assumed that this must be the route by which magnetic information reached the bird's brain. This was borne out by the fact that pigeons trained to detect a strong magnetic field could no longer do so when the trigeminal nerve was cut.[6] A few years later, it emerged that certain areas in the brain of the European robin responded to a rapidly changing magnetic field, but they were inactive when no field was present. When the trigeminal nerve was cut, the activity in these brain regions was significantly reduced.

In light of these findings, the theory that magnetite particles in birds' beaks were indeed the basis of the receptor mechanism was looking promising. But then in 2012, it was revealed that the magnetite particles supposedly found in the pigeons' beaks had been misidentified. They were something entirely different: immune cells called macrophages.[7] And there were other sources of confusion. Several species of birds that migrate by night can manage perfectly well when their trigeminal nerve is cut,[8] while homing pigeons need their olfactory but not their trigeminal nerve to home successfully.[9] On the other hand, reed warblers cannot compensate for a 600-mile eastward displacement (*see page 172*) if the ophthalmic branch of their trigeminal nerve is severed.[10] And strong magnetic pulses that would mess up a magnetite-based receptor do disturb the orientation of adult (but not juvenile) songbirds that migrate by night.[11]

Henrik Mouritsen believes that "the most likely function of the trigeminal nerve-related magnetic sense" is to detect large-scale changes in magnetic field strength and/or inclination, in order to determine the bird's *approximate* position. But how it actually works remains unclear,[12] and a recent experiment[13] suggests that a *gravitational* sensor in the bird's ear called the lagena may possibly also have a part to play in magnetoreception. The situation is therefore very fluid, and if your head is spinning, I can hardly blame you.

While there is still much uncertainty about the role of magnetite, a consensus is starting to emerge around the magnetic-compass sense.

For many years, it has been known that the ability of newts and birds to use a magnetic compass is dependent on the presence of light. As long ago as 1978, Klaus Schulten[14] suggested that chemical reactions in light-sensitive molecules might lie at the heart of this process. In 2000, a specific molecule was proposed in which these reactions might occur: cryptochrome. Almost overnight, the new theory began to attract serious attention.

Cryptochrome molecules are found in many plants and animals, where they are involved in the control of internal clocks and growth. The putative "light-dependent compass" hypothesis depends on the production of "radical pairs" of electrons within these molecules when stimulated by light.[15]

At the heart of the theory is the notion that these radical pairs will behave differently, depending on how the molecules to which they belong are oriented in relation to the earth's magnetic field. The extremely subtle subatomic processes that result may then give rise to a series of further events—a "signaling cascade"—that eventually triggers the firing of a nerve signal. If enough of these events occur, the animal might then become aware of the state of the surrounding magnetic field.

It may seem strange that a light-dependent compass should be used by the many birds—such as robins—that migrate at night, but the cryptochrome mechanism could apparently operate effectively in very low light conditions. Cryptochromes are present in the eyes of birds and, if the radical-pair theory turns out to be correct, the shape of the earth's magnetic field may be superimposed on the bird's normal visual field—rather like a pilot's "head-up" display. They may actually be able to *see* the shape of the magnetic field around them.

Cluster N

One of the leading figures exploring the radical-pair hypothesis is Peter Hore, a professor of chemistry at the University of Oxford. He has been working for a number of years with Mouritsen, and the two men bring to the subject complementary expertise: Mouritsen as an expert on the behavioral and neurophysiological aspects of animal navigation, and Hore as a chemist with a deep knowledge of the properties of radical-pair reactions.

Hore works in a snug office overlooking the leafy backlots of North Oxford, surrounded by crowded bookcases and heaps of papers. He is gentle, modest, and very careful in the claims he makes. But he has devoted his entire career to radical-pair chemistry, and he is the leading expert on the ways in which radical pairs might (and might not) support a biological compass mechanism.

It is a measure of the interest that the radical-pair hypothesis had aroused that the Defense Advanced Research Projects Agency (DARPA)—a powerful but slightly shadowy arm of the US government—approached Hore a few years ago with the suggestion that they might support his work. Plainly DARPA thinks that radical pairs may one day offer more than just an insight into how animals get around. They may even turn out to be relevant to the development of effective quantum supercomputers that could, in principle, perform operations quite beyond the power of any existing computers. Hore did not look this gift horse in the mouth and, in collaboration with Mouritsen, he submitted an application that was quickly rewarded with a large grant.

Though interest in the subject has grown rapidly, progress has so far been slow, largely because it presents so many practical and theoretical problems. In Hore's view, that is unlikely to change soon, though he hopes that in due course—in collaboration with others—he may be able to devise a "killer experiment" that would have the power either to demolish or confirm the cryptochrome hypothesis.

Mouritsen shares Hore's skepticism about the likelihood of rapid progress. His aim is to put together "a bouquet of evidence" from a variety of different sources:

> Trying to understand this magnetic sense requires understanding all levels from the spin of a single electron all the way to a free-flying bird—that's what fascinates me.

Mouritsen has discovered a region of the bird's brain called "Cluster N," which receives input from the eyes. It is the only part of the brain that is highly active when the bird orients in a magnetic field. More revealing still, when Cluster N is destroyed, the bird loses its magnetic-compass sense, while retaining its ability to use its star and sun compasses.[16] These findings argue strongly that the primary magnetic-compass sensors must be in the bird's eye (rather than the beak).

Work on genetically altered insects is also beginning to yield some answers. Cryptochromes have been shown to play an important part in the detection of magnetic fields in fruit flies.[17] And if a cryptochrome similar to those found in mammals is artificially expressed in the eyes of cockroaches, they can be made to change course by exposing them to a rotating magnetic field.[18]

Most of the key experiments in vertebrate magnetoreception involve caged birds in Emlen funnels and, as we have seen already, this troubles researchers like Anna Gagliardo, who prefer to work with free-flying animals (*see page 177*). Mouritsen agrees that it would, in principle, be good to do experiments on free-flying birds, but he points out that there are many variables that are hard to control outside the laboratory. It may, however, soon be possible to extend to birds the techniques for recording from individual cells in the brains of flying bats developed by Nachum Ulanovsky (*see page 228*), in which case there could well be some exciting developments.

There is one other mechanism that might conceivably be involved in magnetic navigation: electromagnetic induction. Viguier discussed this possibility back in 1882, but it has not received nearly as much attention in recent years as the magnetite and cryptochrome hypotheses. The underlying principle—employed in a dynamo—is that an electric current is "induced" in a conductor when it moves through a magnetic field. Electromagnetic induction is, in fact, the process on which we rely for our supplies of electricity.

It is well established that some fish, including sharks and rays, can detect very weak electromagnetic signals and use these for tracking down their prey. To do so, they employ long, jelly-filled canals—splendidly called the "Ampullae of Lorenzini," after the Italian anatomist who discovered them in the seventeenth century. These connect pores in their skin with sensitive detector organs deep inside their bodies.

It was long believed that electromagnetic induction could only work if an animal was surrounded by a medium in which an electrical circuit could easily be completed. Unlike water, air is a poor electrical conductor, but a terrestrial animal might be able to overcome this problem if the whole electromagnetic circuit were enclosed *within* its body. And as it happens, the semicircular canals of a bird's inner ear are filled with a highly conductive fluid that may just fit the bill.[19]

The recent discovery of a structure containing particles of a magnetic mineral in the hair cells lining a bird's semicircular canals has given the electromagnetic-induction hypothesis a boost.[20] The idea is that an electric current may be induced in the liquid that circulates inside these organs and that the hair cells may be able to pick it up.

Far more uncertainty surrounds electromagnetic induction than the other two hypotheses, but it may deserve a closer look.[21]

∞

Bluefin tuna are among the fastest and most powerful swimmers in the sea, capable of moving through the water almost as fast as a cheetah on dry land. They crisscross the Pacific and Atlantic Oceans between their breeding and foraging areas in a highly predictable fashion.[22] They must be very proficient navigators and perhaps they make use of magnetism.

At dusk and dawn, bluefins perform a strange maneuver known as a "spike dive": they descend rapidly into the depths at a steep angle and then return to the surface. The dives occur about thirty minutes on the "dark side" of sunrise or sunset, when the sun is about six degrees below the horizon.[23]

Strangely enough, there is a translucent window on the top of the bluefin's head, just between its eyes. A hollow tube leads from this porthole to the fish's brain, allowing light to reach photosensitive cells on the surface of its pineal gland, which is unusually well-developed. The tube's alignment means that it would point vertically upward during the ascent stage of a spike dive.

One possibility is that the fish sample the polarized patterns in the twilight sky in order to calibrate a magnetic compass. And during the deeper stages of their dives (which can go down to 600 meters), they may be able to measure the magnetic field intensity of the ocean floor with greater precision than is possible nearer the surface. This process could be relevant to the use of a magnetic "map."

Other members of the tuna family are known to be sensitive to magnetic fields,[24] so there is good reason to think that geomagnetism may play a part in bluefin navigation, but nobody knows for sure.

Chapter 24

THE SEAHORSES
IN OUR HEADS

R ats have played an even more prominent part in animal navigation studies than pigeons, bees, or ants. This is partly because they are easy to care for and do not object (too much) to being handled, but an even more important factor is that they are mammals, and therefore much more like us than birds or insects. This has made them an irresistibly attractive subject of research.

Thanks to tens of thousands of experiments, in which rats have been trained to find their way through ingeniously designed mazes, we know that they—like us—rely heavily on landmarks of various kinds to find their way around. So there is surely no need to invoke any "higher" cognitive processes, let alone the use of maps, to explain their navigational behavior. Or is there?

For the first half of the twentieth century, members of the dominant "behaviorist" school of psychology held tenaciously to the theory that all learned behavior could be explained in so-called stimulus-response (S-R) terms. In fairness, S-R theory can account for many of the things that animals learn to do in a laboratory setting. But hard-line behaviorism has long since fallen out of favor. No scientist now would dismiss the possibility that nonhuman animals may have complex mental or indeed emotional lives. As the great primatologist Frans de Waal has put it:

By attributing all behavior under the sun to a single learning mechanism, behaviorism set up its own downfall. Its dogmatic overreach made it more like a religion than a scientific approach.[1]

Even in the heyday of behaviorism, a few open-minded psychologists dared to question the orthodoxy. Edward Tolman (1886–1959), of the University of California, Berkeley, was one of them. In a famous paper published in 1948, he dared to raise doubts about the adequacy of the S-R theory of animal navigation:

> Learning, according to [the behaviorists], consists in the strengthening of some of these connections and in the weakening of others. According to this "stimulus-response" school, the rat in progressing down the maze is helplessly responding to a succession of external stimuli—sights, sounds, smells, pressures, etc.—impinging upon his external sense organs, plus internal stimuli coming from the viscera and from the skeletal muscles. These external and internal stimuli call out the walkings, runnings, turnings, retracings, smellings, rearings, and the like which appear. The rat's central nervous system, according to this view, may be likened to a complicated telephone switchboard.[2]

To Tolman, however, this mechanistic account seemed hopelessly incomplete. His crucial observation was that rats could use shortcuts to find their way to goals they had previously been trained to reach only by longer, indirect routes. They could also make detours when the path they had learned to follow was blocked. How was that possible? It seemed to him that the rats were somehow working out where the goal was located in *space*, rather than blindly following a *fixed route* determined strictly in accordance with the S-R model. In other words, they seemed to be engaged in some form of other-centered navigation.

Further experiments by Tolman and others led him to conclude that rats spontaneously explored their environments and that in the process they built up what he called "cognitive maps," on which all the places and things that mattered to them were somehow recorded. This claim predictably irked the hard-liners, who tried—with an ingenuity reminiscent of medieval theologians—to explain away his results in purely S-R terms.

Tolman was not the first prominent figure to raise the possibility that nonhuman animals might make use of maps. In the 1920s, a leading German psychologist called Wolfgang Köhler published some puzzling observations he had made while he was holed up in the Canary Islands with his pet dog during the First World War.[3]

When Köhler threw a piece of meat out of the window and then closed it, the dog would stand at the window looking longingly at the food and paw the glass. Not so clever, you might think; but if Köhler then also closed the shutters, thus interrupting the dog's view of the food, it would run out of the door and go around the outside of the building to find it.

It looked as if, once the overwhelming visual spell of the food was broken by the closed shutters, the dog stopped to think and was able to recall the layout of the house and garden. Using this information, it could then find an indirect route to its goal—a route it had *never previously been rewarded to follow*. This observation could not readily be explained in S-R terms. It looked very much as if the dog was using some kind of mental map.

The term "cognitive map" is handy shorthand, but it needs to be treated with caution. Rats and dogs obviously do not have maps inside their heads in any literal sense—any more than we do. They are certainly not supplied with them at birth, and they do not, as it were, stop and unfold them when they want to work out where they are. Tolman was speaking metaphorically and meant that the rat's brain was capable of storing geographical information in some kind of code; the analogy of the recently invented digital computer might well have occurred to him.

A cognitive map is best envisaged as a *process*, rather than a thing; a process that emerges from the combined activities of the rat's physically real sensory organs and its central nervous system. That such a process is operating can only be inferred from an animal's behavior—and to make such inferences with any certainty is hard.

Lacking the tools to explore what was actually going on in their brains, neither Tolman nor anyone else in the 1940s could possibly prove that rats (or any other animals) really had "maps" in their heads. But developments in the world of psychology during the 1950s meant that his ideas could be more easily embraced. As the stranglehold of behaviorism gradually weakened, experimental psychologists began to address profound questions—hitherto largely ignored—about how animals and people perceive things, think about them, and solve practical problems.

It became clear that standard S-R learning models could not always provide plausible answers, just as Tolman had argued in relation to maze-running rats. As the great American experimental psychologist, George Miller, succinctly put it: "During the fifties, it became increasingly clear that behavior is simply the evidence, not the subject matter of psychology."[4]

Around the same time, revolutionary technical developments led to the emergence of a whole new discipline: cognitive neuroscience. Very fine wire electrodes inserted into the brains of living animals allowed recordings to be made of the minute electrical signals—only a few ten thousandths of a volt—produced by individual nerve cells (neurons). By patiently making thousands of such recordings, scientists were able to put together a picture of how an animal's brain processes the signals that travel down the optic nerves from its eyes.

They showed that neurons in different parts of the visual cortex were "tuned" to respond to different stimuli. Some, for example, would fire only when the animal was presented with dark bars against a light background, some only in the presence of narrow slits of light against a dark background.[5] It was at last possible to start mapping in detail what different parts of the brain were actually doing.

During the 1950s, the treatment both of severe psychosis and epilepsy often involved removing chunks of the brain. Not surprisingly, these drastic procedures often had unexpected consequences.

One epileptic patient—a young Canadian long known only by his initials "HM," but who deserves to be remembered by his full name, Henry Molaison—was "completely incapacitated by his seizures," and his illness was not responding even to the strongest medication. As a last resort his doctors decided, with his approval, to undertake a "frankly experimental" operation, in which a large section of his two temporal lobes was cut out—including both parts of his hippocampus.[6]

This structure, which is shaped a bit like a seahorse, was named by nineteenth-century anatomists. In the interests of international communication, they used Latin and called it *hippocampus*—a word derived from the Greek for seahorse. Because the brain has two halves (or hemispheres) that closely resemble each other, there are actually two hippocampi: one on either side.

Though Molaison's "understanding and reasoning" were unaffected and his epileptic attacks did become less severe, the operation had "one striking and totally unexpected behavioral result": his memory was drastically impaired. Molaison could no longer recognize the hospital staff or even find the bathroom.

When his family moved to a new house, he was unable to learn the new address and could no longer find his way home, though he still knew how to get to his old house. Molaison could not even recall where objects he used every day were stored and spent hours solving the same jigsaw puzzles again and again. Nor did the crippling memory loss he suffered diminish with the passing of time.[7]

The case of Henry Molaison is famous because it revealed several important things. It provided the first solid evidence that the hippocampus played a key role in memory, and it also became clear that our ability to navigate successfully depended on its integrity. Molaison's sad fate inspired a program of research that has resulted in a series of major advances in our understanding of the neural basis of navigation—and indeed of cognition itself.

The hippocampus lies deep within the brain. Unlike the visual cortex, it is far removed from any direct sensory inputs. Back in the 1960s, most experts doubted whether single-cell recordings from within it would reveal anything comprehensible, let alone that they would shed light on how spatial memories were formed.

Nevertheless, inspired by the case of Henry Molaison, the neuroscientist John O'Keefe (now based at the Sainsbury Wellcome Centre for Neural Circuits and Behaviour in London), assisted by his student, Jonathan Dostrovsky (now at the University of Toronto), decided to explore what was going on in the hippocampus of the rat.

Navigational brain cells

In the early 1970s, O'Keefe's daring paid off when he announced the discovery of individual brain cells that did something extraordinary—in fact, like nothing ever seen before. Each one fired only when the rat occupied a particular point in the cage it was exploring.[8] Or, to put it the other way around, each location the rat visited triggered the firing of a particular cell, or group of cells, in the rat's hippocampus. O'Keefe could actually work out where the animal was simply by looking at the patterns of electrical activity among them.

There was obviously a possibility that the newly discovered cells were firing in response to something else, but nothing the animals could see, smell, or hear had any effect on how the cells behaved. The cells really seemed only to be encoding the spatial properties of the rat's world. O'Keefe therefore decided to call them *place cells*. It was a revolutionary discovery.

In 1978, O'Keefe and Lynn Nadel wrote a book in which they proposed that place cells formed part of an other-centered navigational system that enabled the rat to record and recall the locations of landmarks and goals. In other words, the neurons in the hippocampus were *mapping* the animal's environment. Here, they argued, was the physical basis of Tolman's cognitive map.[9] This was a daring claim at the time, and they certainly aroused the ire of the behaviorists, who were very reluctant to accept their account of what the hippocampus was doing, especially as it seemed to vindicate the views of their old antagonist, Tolman.

Place cells, however, turned out to be only the first in an extraordinary series of discoveries that have over the last fifty years completely transformed how scientists think about the neural basis of navigation—at least in mammals. It is now clear that many different parts of the mammalian brain respond to the spatial properties of the world its owner inhabits, and that successful navigation does not depend exclusively on the hippocampus. The story has therefore become more and more interesting—and complicated.

In the 1980s, another group of cells was found in a brain region right next to the hippocampus called the presubiculum. These fired only when the rat was facing in a particular direction and were accordingly called *head-direction* cells. They responded in exactly the same way, regardless of where the animal was, what it could see, hear, or smell, and whether it was moving or not. They even work in complete darkness and their patterns of firing remain stable over long periods of time. Here then was a group of cells that behaved like a compass, though their activity was not affected by the earth's magnetic field.

More recently, two young researchers at the University of Science and Technology in Trondheim in Norway—Marianne Fyhn and Torkel Hafting—made an even more astonishing discovery. Working under the supervision of the husband and wife team of May-Britt and Edvard Moser, they were investigating cells in an area called the entorhinal cortex (EC) that links the hippocampus to other parts of the brain. They found some that behaved just like place cells, but with one big difference: Instead of firing when the rat was in a single spot, each one of these cells fired in *many different* locations.

This was puzzling, but when they increased the size of the space the rat was allowed to explore, an extraordinary pattern was revealed. It was now clear that the new cells were firing at a series of locations, which formed a regular grid tiling the whole space the rats occupied. These so-called *grid cells* seemed to be recording purely spatial properties of the rat's environment. It was as if the rat was *imposing* a standard grid pattern on the world around it, just as a mapmaker or surveyor might do. They also found head-direction cells in the EC. Some of these formed a grid, too, firing only when the rat visited a particular place and faced in a particular direction.[10]

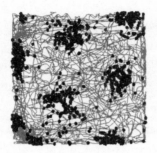

The patterns of firing in a single "grid" cell of a rat exploring a small, square arena. The grey lines show the path the rat follows while the black dots are the "spikes" of electric activity as the animal moves around.

In 2008, the Mosers' team made a further discovery: cells in the EC that fired only when the rat (or mouse) was at the boundary of their cage. They were therefore dubbed *border cells*. Then in 2015, the Mosers revealed other cells that responded only to the rat's running speed—firing more and more frequently as they went faster and faster. In effect, they were acting like a speedometer. The already long list of specialized cells that are involved in navigation continues to lengthen.[11]

These amazing breakthroughs were rewarded in 2014 with a Nobel Prize, which was shared between the Mosers and O'Keefe.[12]

Similar specialized navigational cells have now been found in the brains of mice, monkeys, bats, and people. Opportunities to record directly from single cells in the human brain only arise when electrodes are implanted for medical purposes, but sophisticated brain-imaging techniques now allow scientists to obtain similar results without the need for surgery. The importance of the hippocampus in pigeon navigation is also well established, and although its structure is very different from that of a rat, it also contains specialized "navigational" cells.[13]

But there are still many unanswered questions. Although place cells, grid cells, and head-direction cells may well provide the basis of a "map and compass" system, it is not enough just to know where you are and which way you are heading. You also need to be able to plan a route to your destination and then go there.

Specialized brain cells, which fire as a rat navigates a complex maze, offer a promising clue. These cells, which lie outside the hippocampus, apparently define routes and goals, and others have been found in the hippocampus itself that seem to be involved in route planning.[14]

Obviously, laboratory experiments are highly artificial and do not reflect the reality of life in the wild. The distances involved in real-world navigation may extend to hundreds or even thousands of miles, and while most experiments relate only to navigation in two dimensions, many animals—especially those that can fly or swim—actually need to cope with three. How their brains (and ours) cope with these extremely complex challenges is not yet clear.[15]

It would therefore be very helpful to be able to study what the brain does when an animal is moving freely in its natural environment. An Israeli scientist called Nachum Ulanovsky has, in fact, developed sophisticated methods for recording from single cells in the brains of flying bats,[16] and these may soon be extended to other animals.

While the hippocampus and closely connected areas play a central role in managing navigational tasks, it is evident that other parts of the brain also make important contributions. Signals pass backward and forward between many different brain regions as an animal moves around its environment, as it recalls where it has been, or thinks about where to go next. Exactly how this complex "connectivity" influences navigational processes remains mysterious.

It is also clear that the hippocampus does a lot more than help us map our physical surroundings and find our way around. It is crucial for our memory of people, things, events, and relationships—indeed its basic function may be to provide an abstract "memory space," within which all manner of concepts can be manipulated. On this view, the hippocampus provides the memory bank on which successful navigation depends, rather than actually performing navigational computations.[17]

There is plainly still much we do not know but, in a recent article reviewing the last fifty-odd years of research, the Mosers boldly concluded that navigation might be "one of the first cognitive functions to be understood in mechanistic terms."[18]

However, an interesting philosophical question remains unresolved. Although it is well established that the hippocampus and EC play key roles in navigation, there is room for debate about the basis of the space-time coordinate system they seem to embody. In keeping with classical physics, most neuroscientists take for granted that space and time are fundamental, fixed dimensions of reality—*out there in the world*—that are somehow *represented* in the brain.

But modern physics tells us that space and time are not in fact separate dimensions, and that they are far from being fixed. Our subjective senses of both space and time are also very fluid. So is there another possibility? Perhaps space and time are merely *constructs* that emerge from our physical interactions with the world.[19]

Andreas Pašukonis, a young researcher based at the University of Vienna, has spent a long time in the rainforest of Venezuela patiently studying tiny (one-inch-long) three-striped poison frogs that do something marvelous—and so far unexplained.

The males occupy small patches of the undergrowth, which they defend, and attract females with their calls. After mating, the females lay their eggs, which the males then carefully transport to pools elsewhere in the forest where the tadpoles can hatch and mature. The males then return to their territories. Pašukonis devised a special neoprene jockstrap to attach a radio-tracking device to the males and then transported them to distances of up to 800 meters from their homes.

To Pašukonis's amazement the frogs were not only able to find their way back, but followed quite direct routes, even though their journeys sometimes lasted several days. Given that the rainforest is such a cluttered environment, full of noises, smells, and obstacles, and offering little access to the sky, it is very hard to understand how they do this.[20]

Chapter 25

THE HUMAN NAVIGATIONAL BRAIN

There are many deep questions that remain to be answered, but it is clear why poor Henry Molaison suffered such severe memory loss following the removal of his hippocampi and why, in particular, he had so much difficulty in learning where his new home was. In concert with related brain areas, the hippocampus supports our ability to navigate and provides the basis of Tolman's putative cognitive map.

And it is easy to see why the onset of Alzheimer's is so often heralded by signs of disorientation.[1] The underlying damage frequently appears first in the EC—home of the grid-cell network—before spreading to the hippocampus itself. No wonder then that one of the first questions put to patients with symptoms of dementia is: "Where do you think you are?"

Progress in finding treatments for Alzheimer's (or better still, ways of preventing it) has been slow in coming, but increased knowledge of how the brain enables us to navigate is already helping sufferers to cope better with its disorienting effects. Architects, for example, are now able to plan buildings that patients can navigate more easily.[2] Collaboration between neuroscientists and designers is a growth area, and we all stand to benefit from it, either directly or through improvements to the lives of those who most matter to us.

One of the best-known experiments relating to human navigation involved London taxi drivers, who have to memorize thousands of different routes around the city in order to obtain a license. Getting "The Knowledge," as it is called, is an extremely laborious process that usually takes two or three years to complete, and not everyone passes the final test. Using an MRI brain scanner, Eleanor Maguire and her team showed that the rear part of the hippocampus in taxi drivers was significantly larger than in control subjects.[3]

Moreover, the degree to which its size increased was related to how long they had been driving a taxi—the longer, the bigger.[4] Interestingly, London bus drivers with similar lengths of service did not show the same changes in hippocampal volume, presumably because the task of following the same route day after day is far less navigationally demanding than that faced by taxi drivers.[5]

Maguire's findings imply that the size of the hippocampus is related to the amount of "exercise" it is given—in other words, how often we do things that activate it. If we spend a lot of time using our spatial memory to navigate, we can expect it to grow, and vice versa.[6] In keeping with the "use it or lose it" world view, some researchers have even suggested that we should make a special effort to use our spatial memory as we grow older—rather than relying exclusively on GPS—since that may reduce the risk of developing disorders like Alzheimer's, as well as protecting us from the normal, age-related decline in navigational ability.

This theory has attracted a good deal of media attention, but there does not yet seem to be any direct evidence to support it. I asked Martin Rossor, the UK's national director for Dementia Research, and his colleague, Jason Warren, whether they thought that shrinkage of the hippocampus due to lack of use might increase the likelihood of developing Alzheimer's.

Rossor was cautious. He could see no reason why loss of hippocampal volume would by itself increase the likelihood of developing the disease. Nevertheless, he thought it possible that the "cognitive reserve" of a patient with a relatively small hippocampus would be less than that of someone whose hippocampus was larger.[7] In other words, the severity of the effects of the disease might depend in part on how well-developed the affected parts of the brain were before its onset. So, yes, a person whose hippocampus was small—perhaps through lack of use—might have less resilience in the face of Alzheimer's.

Warren, however, warned that there was a "chicken and egg" problem:

> If you take someone like me who has an appalling sense of direction, I'm going to clutch at any form of electronic assistance I can get, because then there's some vague chance I might find my way from A to B. If I then develop Alzheimer's Disease, is that because I used the electronic assistance, or because I had a weak hippocampal navigation system?

Rossor also pointed out that Alzheimer's is not always associated with navigational difficulties. It all depends on where the plaques and tangles that are the defining characteristics of this condition actually develop in the brain. And problems in finding your way may also reflect difficulties that have nothing to do with navigation. For example, in some forms of dementia, people lose the ability to *recognize* places. They may know that they are in a hospital and even how they got there, but their inability to name the building makes it look as if they are lost. And, more straightforwardly, if someone cannot say where they are, they may just have forgotten how they got there.

Conceptual navigation

We talk about "being on top" of the world or "going downhill," we "look into things" and have "close friends" or "distant relations." Thomas Kuhn, the great philosopher of science, described scientific theories as "maps," and people often talk of "mapping" their relationships. Human language relies heavily on spatial metaphors, and we constantly employ them both in conversation and in our thought processes—and that is probably no accident. It may well reveal something profound about how our minds work.

One of the most fascinating theories to emerge from the world of neuroscience is that the parts of the human brain that support *geographical* navigation, including in particular the hippocampus, may also be involved in *conceptual* navigation.[8] It was long believed that our "higher level" thought processes and our wonderfully flexible intelligence depended on the workings of the prefrontal cortex, but we now know that it cannot manage on its own. Such diverse activities as conducting a conversation, managing social relationships, making sensible decisions, manipulating ideas, making plans for the future, and even exercising our creativity are impossible without a healthy hippocampus.[9]

Our complex social structures probably owe a lot to our ability to plot the positions of our fellow creatures both in physical and conceptual space, and to make accurate predictions about their likely future behavior. It is a striking fact that people of both sexes can more accurately estimate the position of a person than an inanimate object,[10] and there is some evidence that rats, mice, and bats have specialized brain cells that serve to track the position of other members of the same species.[11] Our ability to empathize with other people may also depend on the integrity of the hippocampus.[12]

In a fascinating recent experiment,[13] eighteen people took part in a role-playing game while their hippocampi were monitored in a brain scanner. In the game, the participants moved to a new town and had to find a job and somewhere to live by getting to know the town's people. They were shown slides of cartoon characters "speaking" through word bubbles. The outcome of each interaction reflected changes in the relationships between the participants and the fictional characters.

The concurrent changes in their hippocampal activity suggested that the participants were navigating "a social space framed by power and affiliation." The authors concluded that the concept of social space is more than a mere metaphor; it may indeed "reflect how the brain represents our position in the social world."

This all makes good sense from an evolutionary point of view. Our hunter-gatherer ancestors obviously needed to know and remember where resources of game, edible plants, and water could be found, but it was also vitally important for them to keep track of their *relationships* with other members of the tribe—whether family, friends, allies, foes, or mates.

Recent research with Namibian tribespeople even suggests that male superiority in navigational tasks may be an evolutionary consequence of the fact that men who traveled farther to find sexual partners had more offspring than their competitors.[14] It is not an exaggeration to say that our very lives depend on the ability to make use of mental maps that record not just places, but relationships.

Knowing where we are, as well as the changing locations of other people, animals, and things, and our relationships with them, are vital ingredients of our physical, social, and cultural lives. But so is the ability to think creatively and to place ourselves in imaginary future situations.

Anyone who tries to define the meaning of such a notoriously slippery term as creativity is asking for trouble, but the coupling of images and ideas to produce something entirely novel surely captures important aspects of it. These activities closely mirror what we do when we plan a new route in our heads. Although other parts of the brain—notably the prefrontal cortex—are known to play a key role in creative thinking, researchers have recently shown that "creativity" also depends on having a healthy hippocampus.[15]

In one test, participants are asked to find ways of making a toy more fun to play with, to come up with new uses for a cardboard box, or to make novel drawings starting with only the outline of an oval-shaped figure. Patients who have suffered severe hippocampal damage and associated memory loss, but who have no other cognitive problems, score worse than healthy subjects.

They struggle to generate new ideas, and those they produce are judged to be less novel and interesting than those of controls with no such damage. And it is the same when they are faced with three-word lists composed of words related to a "target word" (for "cream," "skate," and "water," the target word would be "ice"). They find it much harder to identify the target words than healthy subjects.[16]

One final piece of research provides rather more direct evidence that conceptual and spatial navigation in humans depend on similar brain processes. The characteristic patterns of firing found in the grid cells that support map-like representations of space also appear when human subjects perform an entirely abstract cognitive task that has nothing whatever to do with navigation.[17]

These patterns are found not just in the brain areas (like the EC) that are active during physical navigation, but also in those (such as the prefrontal cortex) that are known to be involved in applying learned concepts to novel situations. This suggests that our ability to manipulate concepts is based on the same principles as our capacity to record and analyze spatial relationships.

Every week new discoveries are announced, and before long, neuroscientists may well be able to give a more precise and detailed account of the mechanisms that govern both physical and conceptual navigation. But it is already clear that the navigational computer in our heads is not simply an add-on accessory that only fires up when we undertake a physical journey. The brain circuits that enable us to find our way around have a far wider and deeper significance: They play a key role in shaping our lives and defining who we are.

A pioneering online survey has recently been used to explore the navigational abilities of more than 2.5 million people—from all around the world. The participants played a mobile video game called "Sea Hero Quest," which can be downloaded as an app.[18] If the ability to play an online game is a reliable indicator of navigational ability in the real world then the results suggest that it declines steadily with age—regardless of geographical location. It also appears that men are generally more efficient navigators than women, though interestingly the magnitude of the gender differences is closely linked to measures of social inequality.

Perhaps, then, women have just the same innate navigational potential as men, but are often unable to realize it because opportunities to practice their skills are more limited. Yet another example of gender bias.

Intriguingly, inhabitants of Nordic countries emerged as the world champions. The authors speculate that the long-standing popularity of orienteering in that part of the world might explain their superior navigational skills, though there is another possible explanation: Perhaps they just play a lot of video games during the long winter nights!

∝⌣

"Elephants never forget"—or so they say—and that piece of folklore seems to have some foundation.

African savannah elephants sometimes travel more than sixty miles to find food or water, and they are very good at working out where other elephants are—even when they are out of sight. Using tracking devices, researchers have shown that they have "remarkable spatial acuity." When finding their way to waterholes, they headed off in exactly the right direction, on one occasion from a distance of roughly thirty miles. What is more, they almost always seem to choose the nearest waterhole. The researchers are convinced that the elephants always know precisely where they are in relation to all the resources they need, and can therefore take shortcuts, as well as following familiar routes.[19]

Although the cues used by African elephants for long-distance navigation are not yet understood, smell may well play a part.

Elephants are very choosy eaters, but until recently little was known about how they selected their food. One possibility was that they merely used their eyes and tried out the plants they found, but that would probably result in a lot of wasted time and energy, not least because their eyesight is actually not very good.

The volatile chemicals produced by plants can be carried a long way, and they are very characteristic: Each plant or tree has its own particular odor signature. What is more, they can be detected even when they are not actually visible. New research suggests that smell is a crucial factor in guiding elephants—and probably other herbivores—to the best food resources.

The researchers first established what kinds of plant the elephants preferred either to eat or avoid when foraging freely. They then set up a "food station" experiment, in which they gave the elephants a series of choices based only on smell. The experiment showed that elephants may well use smell to identify patches of trees that are good to eat, and secondly to assess the quality of the trees within each patch. Free-ranging elephants presumably also use this information to locate their preferred food.[20]

Their well-developed hippocampal structures may enable elephants, like rats and people, to construct cognitive maps.

PART 3

WHY DOES NAVIGATION MATTER?

Chapter 26

THE LANGUAGE
OF THE EARTH

Having almost miraculously survived a year of horrors in the concentration camp at Auschwitz, Primo Levi (1919–87), the famous Italian writer and chemist, was too weak and ill to return directly to his home in Turin. At one stage in his long homeward journey, he spent two months convalescing at a camp in the former Soviet Union. The forest surrounding it exercised a powerful attraction on him, as well as his fellow inmates:[1]

> Perhaps it offered the inestimable gift of solitude to all who sought it; we had been deprived of this for so long! Perhaps because it reminded us of other woods, other solitudes of our previous existence; or perhaps, on the other hand, because it was solemn and austere and untouched like no other scenery known to us.

A short distance from the camp, the woods closed in and every trace of animal life disappeared:

> The first time I penetrated it, I learnt to my cost, with surprise and fear, that the risk of "losing oneself in a wood" existed not only in fairytales. I had been walking for about an hour, orienting myself as best I could by the sun, which was visible occasionally, where the branches were less thick; but then the sky clouded

over, threatening rain, and when I wanted to return, I realized I had lost the north. Moss on the tree trunks? It covered them on all sides. I set out in what seemed to be the correct direction; but after a long and painful walk through the brambles and undergrowth, I found myself in as unrecognisable a spot as that from which I had started.

After several hours stumbling through the forest, Levi was convinced that he would die there:

> I walked on for hours, increasingly tired and uneasy, almost until dusk; and I was already beginning to think that even if my companions came to search for me, they would not find me, or would only find me days later, exhausted by hunger, perhaps already dead . . . Then I decided to set off straight ahead, generally speaking towards the north (that is, leaving on my left a slightly more luminous bit of sky, which should have corresponded to the west), and to walk without stopping until I met the main road, or in any case a path or a track.
>
> So I continued in the prolonged twilight of the northern summer, until it was almost night, a prey now to utter panic, to the age-old fear of the dark, the forest and the unknown. Despite my weariness, I felt a violent impulse to rush headlong in any direction, and to continue running so long as my strength and breath lasted.

"Woods shock" is the evocative name given to the terror-stricken state described by Levi. This kind of disorientation is a waking nightmare, in which nothing seems to make sense and everything takes on a sinister aspect. The world itself becomes uncanny—beyond our "ken" or knowledge—and therefore threatening. We literally do not know which way to turn. In this condition, the risk of making life-threatening mistakes is greatly increased.

Eventually Levi heard a distant train whistle and realized that he had been heading in completely the wrong direction. He found his way to the railway and followed it northward by keeping his eye on the Little Bear, the constellation that includes Polaris, which had luckily just appeared through the clouds.

Not many of us now, if caught in a similar predicament, would know how to do that. And people with the exceptional field skills of Enos Mills—the mountain guide who survived after becoming snow-blind when alone in the Rocky Mountains (*see pages 92–93*)—are now rare enough to be regarded as prodigies.

According to Rebecca Solnit, who has interviewed search-and-rescue teams in the American wilderness:

> . . . a lot of the people who get lost aren't paying attention when they do so, don't know what to do when they realize they don't know how to return, or don't admit they don't know. There's an art of attending to weather, to the route you take, to the landmarks along way, to how if you turn around you can see how different the journey back looks from the journey out, to reading the sun and moon and stars to orient yourself, to the direction of running water, to the thousand things that make the wild a text that can be read by the literate. The lost are often illiterate in this language that is the language of the earth itself, or don't stop to read it.[2]

Most of us city-dwellers have quite simply lost the age-old habit of closely observing our surroundings and of constantly—even if unconsciously—keeping track of where we are and which way we are headed. Instead we rely on electronic gadgets to find our way around. This normally presents few problems, but batteries can run down and the satellite signal can be easily lost, or even jammed. The latter is a serious but little-discussed threat.

The signal from a GPS satellite is very weak—in fact, no stronger than a car headlight. Since the satellites are 12,550 miles above the earth's surface, it is all too easy to jam the signals by transmitting a more powerful one on the same frequency. And jamming devices designed to do exactly that are readily available on the internet. They are used by criminals to conceal the movement of vehicles fitted with tracking devices, and they can interfere with GPS receivers over quite a wide radius. Have you sometimes lost the GPS signal for no obvious reason? You may have been the victim of jamming without even knowing it.

There is also the threat of "spoofing," the deliberate transmission of a signal that pretends to come from a GPS satellite but is actually designed to make your receiver give you the wrong position. It is a proven technology that has already caused problems for ships near the coasts of North Korea and Russia. Like jamming, it could be used as a potent weapon of war or terrorism.

But there are much deeper problems. According to Nicholas Carr, an author who has studied our dependence on automatic systems, computers make us vulnerable to cognitive errors of two kinds:

> Automation complacency occurs when a computer lulls us into a false sense of security. Confident that the machine will work flawlessly and handle any problem that crops up, we allow our attention to drift. We become disengaged from our work, and our awareness of what's going on around us fades. Automation bias occurs when we place too much faith in the accuracy of the information coming through our monitors. Our trust in the software becomes so strong that we ignore or discount other information sources, including our own eyes and ears. When a computer provides incorrect or insufficient data, we remain oblivious to the error.[3]

Sometimes these errors have ludicrous consequences; people have been known to drive into rivers while blindly following the directions issuing from their GPS. But they can also lead to disasters, such as plane crashes and shipwrecks.[4] There is also the danger that people will misuse technology. Satnav systems designed for road use should not be used when hiking on a mountain or sailing a boat, yet many people make those errors. Ruth Crosby, of the Loch Lomond National Park in Scotland, says she and her colleagues often get asked for the postcode of Ben Lomond by hikers who plan to climb the mountain armed only with a cell phone.[5]

Some unfortunate people never acquire even the most basic navigational skills. Their brains seem to be in perfectly good shape—unlike those of Alzheimer's patients—but they quickly get lost even in areas they have known for many years. The first such case was described in 2009 and the condition was named "developmental topographical disorientation" (DTD).

Since then more than a hundred other examples have been identified with the help of an online survey. Subsequent tests confirmed that these DTD sufferers—85 percent of whom were women—were far worse than controls at tasks involving orientation, landmark recognition, and retracing their steps, though they were just as good at face and object recognition. It is not yet clear what causes DTD, which seems to be a lifelong affliction, or whether women really are more prone to it than men; perhaps they are simply more willing to admit to having the problem.[6]

DTD sufferers have no choice in the matter, but most of us now routinely make journeys without really grasping where we have been or how we got there. We are transported like parcels, happy to arrive safely at our chosen destination and relieved if our journey has been trouble-free. Modern travel encourages passivity; we are only too willing to leave the navigation to others, whether the pilot of the aircraft, or the beguilingly confident voice of the GPS that accompanies us everywhere. Self-driving cars are taking this dependency to a new level.

Our long-dead ancestors explored almost the entire surface of our planet and colonized large parts of it without the help of any tools, apart from their finely tuned senses and native wits. Long before the magnetic compass, astrolabe, sextant, and marine chronometer were invented—let alone GPS—they had, as we have seen, developed an astonishing variety of wayfinding skills adapted to environments ranging from the high Arctic to the deserts of Australia and the tropical waters of the Pacific Ocean.

The following anecdote, recounted to Claudio Aporta by an Inuit elder called Ikummaq in 2000, reveals how modern technology is threatening those old ways:

> If a young person asks a GPS where a certain place is, that GPS will tell him, tell that young person. But if that young person approaches an elder and asks where that certain place is, that elder is going to go detail by detail, and then describe what is before that, not necessarily where that is. Describe "this comes out first, like a bay, a point, an inukshuk," and so on and so forth. As you progress, they'll tell you exactly what to expect. And a youth doesn't have time for that. He wants to know where that place is . . .
>
> There are some people of my age who rely on GPS because their fathers didn't sit down with them or take them out on the land to teach them where to go, how to get there, what's dangerous. They haven't done that. And over time, if you keep practicing, [Inuit navigation] is almost like a science. And maybe it is a science, as a matter of fact, but nothing written. It's just mental, it's just knowledge passed on from generation to generation.[7]

Although satellite navigation offers many practical advantages, its adoption has led to a deterioration in wayfinding skills and, more generally, "a weakened feel for the land":

> An Inuit on a GPS-equipped snowmobile is not so different from a suburban commuter in a GPS-equipped SUV: as he devotes his attention to the instructions coming from the computer, he loses sight of his surroundings . . . A unique talent that has distinguished a people for centuries may evaporate in a generation.[8]

Bob Gill, of the US Geological Survey in Anchorage, who discovered the extraordinary feats of endurance of the bar-tailed godwit, has told me that the native people of Alaska have their own name for GPS.

They simply call it "the elders in a box."

Though the ancient skills of the Pacific Islanders have been revived, elsewhere the old ways are in serious jeopardy and may soon survive only in myth and legend. Their loss would sever one of the most important surviving links between us and our not-so-distant, hunter-gatherer ancestors. The GPS revolution is the latest stage in a long historical process that has seen us abandon—one after another—most of the practical skills on which our ancestors once depended. We are only too happy to leave it to specialists to grow our food, make our clothes, and build our homes. And now we are turning our back on perhaps the oldest and most fundamental skill of all: navigation.

Asked how he went bankrupt, a character in one of Ernest Hemingway's novels[9] answers: "Gradually and then suddenly." The loss of our navigational skills has happened in much the same way. It began, slowly, with the adoption of earlier, simpler technologies like the compass and sextant, but these did not relieve us of the need to pay close attention to the world around us and to use our wits.

The arrival of GPS has, by contrast, brought about an abrupt and fundamental change in our relationship with nature. Now we can fix our position and set a course without the slightest thought or effort—without so much as raising our eyes from our glowing screens. The gadgets that seem to have relieved us of a tiresome burden are not only enfeebling us but also distancing us from the natural world.

GPS is almost miraculous, and it stands as one of the greatest technological achievements of modern times. But in our devotion to it, are we perhaps behaving a little bit like Faust—the man who sold his soul in return for the granting of his dearest wishes? Though we may not realize it, we are fast becoming navigational idiots. To avoid that fate, we need to put aside our cell phones and electronic navigation systems whenever we can. Rather than automatically relying on GPS, even when following a route we know perfectly well, we should open our eyes and exercise our brains. Unless we want to lose our navigational skills altogether, we must learn again how to speak the language of the earth.

＊

Geraldine Largay, a sixty-six-year-old retired nurse, set off from Harpers Ferry in West Virginia on April 23, 2013, with her traveling companion, Jane Lee. Their ambitious plan was to hike all the way to the northern end of the Appalachian Trail, a distance of some 1,100 miles.

Lee had to return home at the end of June, but Largay was determined to carry on alone, even though she had a poor sense of direction and suffered from anxiety attacks. On July 21, she was still going strong and was within 200 miles of the end of the trail at Mount Katahdin in Maine; by now she had covered nearly 1,000 miles. The following day she had a rendezvous with her husband, who was bringing her supplies for the next leg. Another hiker took a picture of Largay at about 6:30 AM on July 22, just as she was setting off. She was the last person to see Largay alive.

On July 24, Largay's husband reported that she was overdue and the Maine Warden Service began to search the heavily wooded mountains around her last reported location. Many other agencies joined the massive hunt, with the help of aircraft and sniffer dogs. Although the initial search was called off after a week, the case remained open and the police continued to chase up a series of spurious leads. In October 2015—more than two years later—a surveyor happened upon a collapsed tent, in which he found Largay's remains.

Largay's mobile phone revealed that she had left the trail on the morning of July 22 to relieve herself. Disoriented and unable to regain the path, she tried repeatedly to send text messages to her husband, but because the cellular coverage was poor or nonexistent in this remote and mountainous corner of Maine, he received none of them.

Her final camp was only two miles away from the trail, and search parties came close to her location on more than one occasion. The entries made in a journal found in the tent revealed that she had survived at least until mid-August 2013, by which time her supplies had run out. The last clearly dated entry was on August 6, 2013. It was forlorn but serene:

When you find my body, please call my husband George and my daughter Kerry. It will be the greatest kindness for them to know that I am dead and where you found me—no matter how many years from now.[10]

Chapter 27

CONCLUSIONS

The North American monarch butterfly is in decline, and the scientists who study its migratory behavior are helping to find out why. The reasons include the destruction of the highland forests in which they pass the winters, and the widespread use of glyphosate herbicides (like Roundup) in the Great Plains of the US, which are killing off the plants on which their larvae depend. If effective steps are not taken to deal with these threats, an annual event that must count as one of the most impressive of all the natural phenomena may soon be just a memory.

We know that glyphosate weakens the navigational abilities of honeybees,[1] and it may well be implicated in their decline—a problem that seriously threatens agricultural productivity, because of the vital role these insects play as pollinators. The dangers of herbicide use almost certainly extend to many other insect species.

Habitat loss is endangering countless animals, and migratory birds are especially at risk. The heroic bar-tailed godwit, for example, on its return trip from New Zealand to Alaska must stop to refuel in wetlands on the coast of China, and because these are shrinking fast, its survival is in question. Changes in the circulation of the great ocean currents and wind systems are likely consequences of climate change, and they will seriously threaten the many animals—from turtles and whales to arctic terns and dragonflies—that depend on them.

We know that light pollution is a serious threat to many animals. Artificial lights lure hatchling turtles away from the sea and dangerously confuse many species of birds and insects. They have a disastrous effect on the internal clocks that govern the navigational behavior of many animals. Tackling this growing and largely unnecessary problem is a major challenge and one that is still not widely enough appreciated.[2]

I could go on, but even these few examples show how animal navigation studies are informing efforts to conserve the breathtaking array of creatures, great and small, with which we share this planet—and to combat environmental change.

From an entirely selfish, human perspective, understanding the factors that govern the movements of agricultural pests, such as locusts and cutworm moths (including the bogong), is of great economic and social value. And controlling the spread of dangerous diseases (like influenza and malaria) that are carried by animals depends on knowing where they go, when, and why. These are all problems to which animal navigation scientists have made, and continue to make, vital contributions.

Thanks to the work of neuroscientists, we know that exercising our navigational skills may help us cope better with the normal age-related decline in our navigational abilities, and even perhaps the destructive inroads of Alzheimer's disease. Knowledge about how the brain manages navigational tasks can also help us support victims of Alzheimer's more effectively; for example, by designing environments they can navigate with greater ease and safety.

Our growing understanding of the sensory and computational processes that underlie human and animal navigation is already shaping the development of radical new technologies. From self-driving vehicles and robotic systems to machine vision and even perhaps quantum computing, these have the power to transform the world we live in. Such developments have many potential applications in the military and security arenas, which helps to explain why a good deal of the funding for animal navigation research comes from government sources. Whether we use our new knowledge for good or evil is up to us.

Each of us follows a path through time and space—a lifeline, if you like—that shapes the story of our lives. When we wake from a deep sleep, our ability to remember who we are depends on recalling where we have been, who we have encountered, what we have done and where. These are the things that give us a sense of enduring personal identity and without them our lives fall apart completely—as we see in advanced cases of Alzheimer's disease. By shedding light on how we construct our sense of the self, the neuroscience of navigation is helping us to understand who we are, and how much we have in common with our animal cousins.

We humans have long prided ourselves (in the Western world at least) on our superiority to the rest of "creation." Our special status is enshrined in the Book of Genesis, where it is proclaimed that God "created man in his own image" and gave him "dominion over the fish of the sea, over the fowl of the air, and over every living thing that moveth upon the earth." St. Augustine went much further. He argued that we had no moral duty toward other animals—citing as evidence the fact that Jesus had cast devils out of a man and sent them into a herd of swine, which he then allowed to drown.[3] Our fellow creatures therefore existed only to be used by us and their welfare was of no intrinsic importance.

In the middle ages, St. Thomas Aquinas[4] adopted a more moderate position, suggesting that we should be kind to animals, because we might otherwise acquire habits of cruelty that could spill over into our treatment of people. But he did not question our fundamental superiority as humans. And it was not only Christian authors who embraced anthropocentrism. Aristotle maintained that nature had made all things specifically for the sake of man.[5]

The Darwinian revolution posed a devastating challenge to this deeply anthropocentric worldview, and subsequent scientific advances have demolished its intellectual credibility. We may in some respects be more gifted than our fellow creatures, but in others they are plainly superior to us. The key point is that the differences, in both directions, are more of degree than of kind.

Humans do not belong to a different order of being; we are animals, too, and are the product of the same evolutionary processes that have given rise to bacteria, jellyfish, centipedes, lobsters, birds, and elephants. What sets us apart is that we are in a position to influence the fate of every other creature on the planet—and we have some choice in the matter.

Old habits of thought (and belief systems) die hard, and anthropocentrism remains deeply embedded in our ways of thinking. In fact, it continues to exert a powerful influence on public life, notably in the US, where fundamentalist religious ideas underlie the refusal of many politicians to accept that climate change is a reality.[6] But the problems run much deeper than that. Those who regard Biblical revelation as a more reliable source of information about the world than science have little hope of understanding, let alone solving, the many practical problems that face us. Skepticism about science licensed by religious belief is allowing our "leaders" to deride "expert opinion," when it presents an inconvenient challenge to their ill-informed and sometimes dangerous opinions.

Anthropocentrism not only weakens our capacity to respond intelligently to the dangers we face, but it also gives us an excuse to treat the entire natural world with contempt. I am not talking only of the ill-treatment of millions of farm animals, though that is bad enough. We are rapidly destroying whole ecosystems—from the melting Arctic and the bleached coral reefs of the tropics, to the logged-out rainforests of the Pacific Northwest and the overfished oceans. We are witnessing (and indeed causing) a biological holocaust that would be appalling even if it did not also represent a real threat to our own welfare.

Anthropocentrism is a destructive and dangerous force, and one that we must overcome if we are to take the steps needed to limit the damage we are causing to the world we live in. That is not going to be an easy task, not least because we humans are far from being entirely rational beings. We are all subject to powerful social pressures and prefer to conform with those whose opinions matter to us. We tend to ignore any evidence that threatens our existing beliefs and to seize on any that supports them, and we often jump to conclusions before examining all the evidence carefully.

If progress is to be made in meeting the many environmental problems we face, we must not only challenge the skeptics but also encourage those who recognize the need for change, yet are afraid to take the politically difficult steps that are so urgently needed. We may make faster progress if we avoid focusing exclusively on gloomy prognostications about what the future holds. There is a danger that, by encouraging fatalism, such prophecies may actually prove self-fulfilling.

It is more important to remind ourselves that we live amidst wonders and to widen as far as possible the circle of those who appreciate how remarkable our fellow creatures really are. It would be absurd to pretend that discoveries about animal navigation are by themselves going to make a big difference, but they can help us to recognize the value of all that is at risk.

Our species has been around for 300,000 years, and we have spent at most 10,000 years living in villages or towns. Cities with more than a million inhabitants have existed only for a few hundred years, but now most of us are crowded into them, largely cut off from nature—except in the form of parks and the few trees, plants, and animals that can tolerate urban life alongside us. A fundamental feature of the lives of our ancestors was their immersion in the natural world, but for the vast majority of people that is no longer even a memory.

In evolutionary terms, the radical shift from a hunter-gatherer existence to predominantly urban ways of life has happened in the blink of an eye. Whether we like it or not, the deep past still exerts a profound influence over us—through our genes and through the cultures to which we belong—and there is no doubt that the natural world is still of vital importance to us. The great entomologist E. O. Wilson believes that we have inherited an "urge to affiliate with other forms of life" and he has given it a name: "biophilia."[7]

We do indeed seem to be drawn to "nature," in all its wonderfully various forms. Some of us may love hiking in the mountains, while others prefer fishing beside a quiet stretch of river, or sailing on the open sea. But whatever our personal preference may be, there is a wealth of evidence that contact with the natural world is not only enjoyable, but good for us.

In fact, a dose of nature can sometimes prove transformational. Victims of war who have retreated into silence may learn to live again after a couple weeks kayaking down the rapids of the Colorado River.[8] Even the sight of a garden from a hospital window helps patients recover faster from surgery, and taking long walks in the woods (a therapy known in Japan as *shinrin-yoku*, or "forest bathing") reduces stress, as well as producing a host of other beneficial effects.

There are many such examples in medical literature. Improved functioning of the immune system is thought to be one of the underlying mechanisms.[9] There is even evidence that the experience of "awe" elicited by natural phenomena encourages us to behave better—to be less selfish and more cooperative.[10]

The obvious benefits of city life and modern technology, in the end, cannot compensate us for the loss of something mysterious that physical contact with nature alone seems able to supply. Perhaps we are so strongly drawn to the natural world because it is, in some deep sense, our real home—and we long to return to it.

Nature can be overpowering and sublime. Think of the ancient layered cliffs that tower above the Grand Canyon, the dazzling array of stars in a dark night sky, or the endless vistas of the open ocean. The grandeur of such spectacles offers a silent rebuff to our cheeky self-importance. But little things can touch us just as deeply: the sharp dips and turns of a swallow hunting furiously for insects to fuel its long autumn journey; a dung beetle trundling its ball across the hills of Provence; a turtle diligently laying her eggs on a tropical beach; the green-glowing trail of a billion plankton sparkling in the wake of a ship as it sails through the night; or millions of small brown moths setting their course by the magnetic field that envelops the earth.

In researching and writing this book, I have again and again been struck dumb with admiration by the extraordinary skills of the animal navigators that are its stars. Even if our own lives did not depend on the health and vitality of the planet we inhabit, the preservation of the almost infinitely complex web of life from which such wonders emerge is surely an ethical imperative.

The sense of awe we feel in the presence of nature is a mysterious force. At one time it was taken as a sure sign of a divine presence. We may no longer believe in gods, but if we are to flourish, we must learn to respect and care for the world we inhabit and the extraordinary creatures with which we share it.

We must plot a new course.

Acknowledgments

First, I would like to thank my agent, Catherine Clarke, and Rupert Lancaster, my editor. Catherine patiently helped me to develop the original proposal, and Rupert's good-humored but expert advice has played a crucial part in shaping the book. I am also very grateful to the copyeditor, Barry Johnston; the illustrator, Neil Gower; the publicist Karen Geary and her assistant Jeannelle Brew; Caitriona Horne, who managed the marketing campaign; and Cameron Myers, who pulled all the strands together.

In researching this book I have relied mainly on articles published in scientific journals, but I must also acknowledge my debt to the authors of certain books (listed in the bibliography) on which I have drawn heavily: Hugh Dingle, Paul Dudchenko, James Gould and Carol Grant Gould, Tania Munz, and Gilbert Walbauer.

I am extremely grateful to all the scientists who generously shared their expertise with me: Andrea Adden, Susanne Åkesson, Emily Baird, Vanessa Bézy, Roger Brothers, Jason Chapman, Nikita Chernetsov, Marie Dacke, Michael Dickinson, David Dreyer, Barrie Frost, Anna Gagliardo, Bob Gill, Dominic Giunchi, Anya Günther, Jon Hagstrum, Lucy Hawkes, Stanley Heinze, Peter Hore, Miriam Liedvogel, Lucia Jacobs, Kate Jeffery, Basil El Jundi, Ken Lohmann, Paolo Luschi, Henrik Mouritsen, Martin Rossor, Hugo Spiers, Eric Warrant, Jason Warren, Rüdiger Wehner, and Matthew Witt. I am especially grateful to those who kindly read through the first draft of the book (in whole or in part) and commented on it, often in considerable detail: Jason Chapman, Anna Gagliardo, Peter Hore, Kate Jeffery, Paolo Luschi, Henrik Mouritsen, Martin Rossor, Eric

Warrant, and Rüdiger Wehner. My thanks also go to the "civilians" who read the manuscript and commented on it: Jessie Lane, George Lloyd-Roberts, Richard Morgan, and Kit Rogers.

Eric Warrant very kindly allowed me to join him and the rest of his team in the Snowy Mountains, where I witnessed their fascinating experiments on the bogong moth. He and his wife Sarah showed me great hospitality during my time in Lund, as did Rüdiger Wehner and his wife Sibylle when I visited them in Zürich; Paolo Luschi and his wife Cristina were equally kind to me when I was in Pisa. Vanessa Bézy, Roger Brothers, and Ken Lohmann also looked after me very well when I spent time with them in Costa Rica. I greatly appreciate the kindness of them all. I would also like to express my thanks to the Royal Institute of Navigation (RIN) and its present Director, John Pottle, as well as his predecessor, Peter Chapman-Andrews. The Animal Navigation Conference of the RIN held in 2016 provided me with an excellent overview of the latest research, especially on magnetic navigation, and also enabled me to establish contact with many of the leading scientists working in the field. I also gained many useful insights when I attended a conference on animal navigation organized by the Association for the Study of Animal Behaviour later the same year.

Finally, my deepest thanks go to my wife, Mary, for her constant support, advice, and encouragement, and to my daughters, Nell and Miranda. Their help has been more important than I can say.

Selected Bibliography

Ackerman, J., *The Genius of Birds*, London: Corsair, 2016.

Bagnold, R. A., *Libyan Sands*, London: Eland Publishing, 2010.

Balcombe, J., *What a Fish Knows*, London: Oneworld, 2016.

Cambefort, Y., *Les Incroyables Histoires Naturelles de Jean-Henri Fabre*, Paris: Grund, 2014.

Carr, A., *The Sea Turtle*, Austin, TX: University of Texas, 1986.

Cheshire, J., and Uberti, O., *Where the Animals Go*, London: Particular Books, 2016.

Cronin, T. W.; Johnsen, S.; Marshall, N. J.; and Warrant, E. J., *Visual Ecology*, Princeton: Princeton University Press, 2014.

Deutscher, G., *Through the Language Glass*, London: Arrow Books, 2010.

De Waal, F., *Are We Smart Enough to Know How Smart Animals Are?*, London: Granta, 2016.

Dingle, H., *Migration: The Biology of Life on the Move* (second ed.), Oxford: Oxford University Press, 2014.

Dudchenko, P. A., *Why People Get Lost*, Oxford: Oxford University Press, 2010.

Ellard, C., *You Are Here*, New York: Anchor Books, 2009.

Elphick, J., *Atlas of Bird Migration*, Buffalo, NY: Firefly Books, 2011.

Fabre, J. H., *Souvenirs Entomologiques*, Paris: Librairie Ch. Delagrave, 1882.

Finney, B., *Sailing in the Wake of the Ancestors*, Honolulu: Bishop Museum Press, 2003.

Gatty, H., *Finding Your Way Without Map or Compass*, New York: Dover, 1999.

Gazzaniga, M. S.; Ivry, R. B.; and Mangun, G. R.; *Cognitive Neuroscience: The Biology of the Mind* (second ed.), New York: Norton, 2002.

Ghione, S., *Turtle Island: A Journey to Britain's Oddest Colony*, London: Penguin 2002. Tr. Martin McLaughlin.

Gladwin, T., *East Is a Big Bird*, Cambridge, MA: Harvard University Press, 1970.

Gould, J. L., and Gould, C. G., *Nature's Compass: The Mystery of Animal Navigation*, Princeton: Princeton University Press, 2012.

Griffin, D. R., *Animal Minds*, Chicago: University of Chicago Press, 2001.

Heinrich, B., *The Homing Instinct: Meaning and Mystery in Animal Migration*, London: William Collins, 2014.

Hughes, G., *Between the Tides: In Search of Turtles*, Jacana, 2012.

Levi, P., *If This Is a Man; The Truce* (S. Woolf, trans.), London: Abacus, 1987.

Lewis, D., *We, The Navigators* (second ed.), Honolulu: University of Hawaii Press, 1994.

Munz, T., *The Dancing Bees: Karl von Frisch and the Discovery of the Honeybee Language*, Chicago: University of Chicago Press, 2016.

Newton, I., *Bird Migration*, London: W. Collins, 2010

Pyle, R. M., *Chasing Monarchs*, New Haven, CT: Yale University Press, 2014.

Shepherd, G. M., *Neurogastronomy*, New York: Columbia University Press, 2013.

Snyder, G., *The Practice of the Wild*, Berkeley, CA: Counterpoint, 1990.

Solnit, R., *A Field Guide to Getting Lost*, Edinburgh: Canongate Books, 2006.

Strycker, N., *The Thing with Feathers*, New York: Riverhead Books, 2014.

Taylor, E. G. R., *The Haven-Finding Art: A History of Navigation from Odysseus to Captain Cook*, London: Hollis and Carter, 1956.

Thomas, S., *The Last Navigator*, New York: Holt, 1987.

Waldbauer, G., *Millions of Monarchs, Bunches of Beetles: How Bugs Find Strength in Numbers*, Cambridge, MA: Harvard University Press, 2000.

Waterman, T. H., *Animal Navigation*, New York: Scientific American Library, 1989.

Wilson, E. O., *Biophilia*, Cambridge, MA: Harvard University Press, 1984.

Notes

Preface

1. Navigating without maps or instruments is sometimes known as "way-finding," but, for the sake of clarity and simplicity, I have in general avoided using this term.

Chapter 1: Mr. Steadman and the Monarch

1. Santosh, M.; Arai, T.; and Maruyama, S. (2017). "Hadean Earth and primordial continents: the cradle of prebiotic life," *Geoscience Frontiers*, 8(2), p. 309–27.
2. Dodd, M. S.; Papineau, D.; Grenne, T.; Slack, J. F.; Rittner, M.; Pirajno, F.; . . . and Little, C. T. (2017). "Evidence for early life in earth's oldest hydrothermal vent precipitates," *Nature*, 543(7643), p. 60–4.
3. Adler, J. (1976). "The sensing of chemicals by bacteria," *Scientific American*, 234(4), p. 40–7.
4. Blakemore, R. (1975). "Magnetotactic bacteria," *Science*, 190(4212), p. 377–9.
5. Kirkegaard, J. B.; Bouillant, A.; Marron, A.O.; Leptos, K.C.; and Goldstein, R.E. (2016). "Aerotaxis in the closest relatives of animals," *eLife*, 5, e18109.
6. Reid, C. R.; Latty, T.; Dussutour, A.; and Beekman, M. (2012). "Slime mold uses an externalized spatial 'memory' to navigate in complex environments," *Proceedings of the National Academy of Sciences*, 109(43), p. 17490–4.
7. Tero, A.; Takagi, S.; Saigusa, T.; Ito, K.; Bebber, D. P.; Fricker, M. D.; . . . and Nakagaki, T. (2010). "Rules for biologically inspired adaptive network design," *Science*, 327(5964), p. 439–42.
8. Last, K. S.; Hobbs, L.; Berge, J.; Brierley, A. S.; and Cottier, F. (2016). "Moonlight drives ocean-scale mass vertical migration of zooplankton during the Arctic winter," *Current Biology*, 26(2), p. 244–51.
9. Häfker, N. S.; Meyer, B.; Last, K. S.; Pond, D. W.; Hüppe, L.; and Teschke, M. (2017). "Circadian Clock Involvement in Zooplankton Diel Vertical Migration," *Current Biology*, 27(14), p. 2194–201.

10. Vidal-Gadea, A.; Ward, K.; Beron, C.; Ghorashian, N.; Gokce, S.; Russell, J.; . . . and Pierce-Shimomura, J. (2015). "Magnetosensitive neurons mediate geomagnetic orientation in Caenorhabditis elegans," *eLife*, 4, e07493.

11. Phillips, J., and Borland, S. C. (1994). "Use of a specialized magnetoreception system for homing by the eastern red-spotted newt Notophthalmus viridescens," *Journal of Experimental Biology*, 188(1), p. 275–91.

12. Garm, A.; Oskarsson, M.; and Nilsson, D. E. (2011). "Box jellyfish use terrestrial visual cues for navigation," *Current Biology*, 21(9), p. 798–803.

13. "Homesick sheepdog walks 240 miles home to Wales after bolting from his new farm in Cumbria." *Daily Telegraph*, April 25, 2016.

14. Hart, V.; Nováková, P.; Malkemper, E. P.; Begall, S.; Hanzal, V., Ježek, M.; . . . and Červený, J. (2013). "Dogs are sensitive to small variations of the earth's magnetic field," *Frontiers in Zoology*, 10(1), p. 80.

Chapter 2: Jim Lovell's Magic Carpet

1. Darwin, C. (1871). *The Descent of Man, and Selection in Relation to Sex* (D. Appleton and Company, New York, 2nd ed., 1875), pt I, p. 619.

2. Shubin, N.; Tabin, C.; and Carroll, S. (2009). "Deep homology and the origins of evolutionary novelty," *Nature*, 457(7231), p. 818.

3. Standing, L. (1973). "Learning 10,000 pictures," *Quarterly Journal of Experimental Psychology*, 25, p. 207–22.

4. Aporta, C.; Higgs, E.; Hakken, D.; Palmer, L.; Palmer, M.; Rundstrom, R.; . . . and Higgs, E. (2005). "Satellite culture: global positioning systems, Inuit wayfinding, and the need for a new account of technology," *Current Anthropology*, 46(5), p. 729–53.

5. W. E. H. Stanner, quoted in: Lewis, D. (1976). "Observations on route finding and spatial orientation among the Aboriginal peoples of the Western Desert region of Central Australia," *Oceania*, 46(4), p. 249–82.

6. Deutscher, G., *Through the Language Glass: Why the World Looks Different in Other Languages* (Arrow Books, 2011), p. 166–7.

7. Ibid., p. 187.

8. Cambefort, Y., *Les Incroyables Histoires Naturelles de Jean-Henri Fabre* (Grund, 2014), p. 20.

9. Fabre, J. H. (1882). *Souvenirs Entomologiques* (Vol. 2) Librairie Ch. Delagrav. p. 137–8

10. Fabre, J. H., Ibid. p. 140–53

11. "Petit Poucet" in Perrault's original French version, and "Hop o' My Thumb" in the common English translation. The small boy used a trail of little pebbles to find the way home when he and his brothers were abandoned by his destitute parents. But everything went wrong when he later used breadcrumbs, which were eaten by birds.

12. Summarized in: Gould, J. L., and Gould, C. G., *Nature's Compass: The Mystery of Animal Navigation* (Princeton University Press, 2012), p. 173–6.

Chapter 3: A Tangled Horror

1. Warrant, E. J.; Kelber, A.; Gislén, A.; Greiner, B.; Ribi, W.; and Wcislo, W. T. (2004). "Nocturnal vision and landmark orientation in a tropical halictid bee," *Current Biology*, 14(15), p. 1309–18.

2. Warrant, E. J. (2008). "Seeing in the dark: vision and visual behavior in nocturnal bees and wasps," *Journal of Experimental Biology*, 211(11), p. 1737–46.

3. De Perera, T. B. (2004). "Spatial parameters encoded in the spatial map of the blind Mexican cave fish, Astyanax fasciatus," *Animal Behavior*, 68(2), p. 291–5.

4. Sheenaja, K. K., and Thomas, K. J. (2011). "Influence of habitat complexity on route learning among different populations of climbing perch (Anabas testudineus Bloch, 1792)," *Marine and Freshwater Behaviour and Physiology*, 44(6), p. 349–58.

5. Cain, P., and Malwal, S. (2002). "Landmark use and development of navigation behaviour in the weakly electric fish Gnathonemus petersii (Mormyridae; Teleostei)," *Journal of Experimental Biology*, 205(24), p. 3915–23.

6. Clarke, D.; Morley, E.; and Robert, D. (2017). "The bee, the flower, and the electric field: electric ecology and aerial electroreception," *Journal of Comparative Physiology A*, 203(9), p. 737–48.

7. Kamil, A. C., and Jones, J. E. (1997). "The seed-storing corvid Clark's nutcracker learns geometric relationships among landmarks," *Nature*, 90, p. 276–9.

8. Bednikoff, P. A., and Balda, R. P., (2014). "Clark's nutcracker spatial memory: The importance of large, structural cues," *Behavioural Processes*, 102, p. 12–17.

9. See: rothschildarchive.org/contact/faqs/rothschilds_and_pigeon_post.

10. Biro, D.; Freeman, R.; Meade, J.; Roberts, S.; and Guilford, T. (2007). "Pigeons combine compass and landmark guidance in familiar route navigation," *Proceedings of the National Academy of Sciences*, 104(18), p. 7471–6.

11. Ibid.

12. Mann, R. P.; Armstrong, C.; Meade, J.; Freeman, R.; Biro, D.; and Guilford, T. (2014). "Landscape complexity influences route-memory formation in navigating pigeons," *Biology Letters*, 10(1), 20130885.

13. Quoted in: ox.ac.uk/news/2014-01-22-hedges-and-edges-help-pigeons-learn-their-way-around.

14. Tsoar, A.; Nathan, R.; Bartan, Y.; Vyssotski, A.; Dell'Omo, G.; and Ulanovsky, N. (2011). "Large-scale navigational map in a mammal," *Proceedings of the National Academy of Sciences*, 108(37), E718–24.

15. DeLuca, W. V.; Woodworth, B. K.; Rimmer, C. C.; Marra, P. P.; Taylor, P. D.; McFarland, K. P.; . . . and Norris, D. R. (2015). "Transoceanic migration by a 12g songbird," *Biology Letters*, 11(4), 20141045.

Chapter 4: Of Desert Warfare and Ants

1. Within the tropics the sun at noon is directly overhead on two days each year, but will normally be either north or south of you.

2. In polar regions it does not rise or set at all for part of the year, either being permanently above the horizon (in high summer), or permanently below it (in the depths of winter).

3. Bagnold, R. A., *Libyan Sands: Travel in a Dead World* (Eland Publishing, 2010), p. 220.

4. Ibid., p. 59. See also: Shaw, W. K. (1943). "Desert Navigation: Some Experiences of the Long Range Desert Group," *Geographical Journal*, p. 253–8.

5. Bagnold, R. A., *Libyan Sands*, p. 171–2.

6. Lubbock, J., *Ants, Bees and Wasps: A Record of Observations on the Habits of the Social Hymenoptera* (D. Appleton and Co., New York, 1882), p. 263–70.

7. The following account is based on Wehner, R. (1990). On the brink of introducing sensory ecology: Felix Santschi (1872–1940), Tabib-en- Neml, *Behavioral Ecology and Sociobiology*, 27(4), p. 295–306.

8. For a fascinating historical account of the early students of ant navigation see: Wehner, R. (2016). "Early ant trajectories: spatial behavior before behaviorism," *Journal of Comparative Physiology A*, 202(4), p. 247–66.

9. Jouventin, P.; and Weimerskirch, H. (1990). "Satellite tracking of wandering albatrosses," *Nature*, 343(6260), p. 746.

Chapter 5: The Dancing Bees

10. For an excellent discussion of von Frisch's life and work, see: Munz, T., *The Dancing Bees: Karl von Frisch and the Discovery of the Honeybee Language* (University of Chicago Press, 2016). I have drawn heavily on her book in this chapter.

11. Ibid., p. 151.

12. Munz, T. *The Dancing Bees*. p. 162–3.

13. Ibid. p. 184–5.

14. Ibid. p. 92.

15. Von Frisch, K.; and Lindauer, M. (1956). "The 'Language' and Orientation of the Honeybee," *Annual Review of Entomology*, vol. 1, p. 45–8.

16. Ibid.

17. The bees must be able to judge which way is "up" in order to interpret the "dance"; something that is only possible in the dark of the nest by sensing the downward pull of gravity.

18. Munz, T. (2005). "The bee battles: Karl von Frisch, Adrian Wenner and the honeybee dance language controversy," *Journal of the History of Biology*, 38(3), p. 535–70.

19. Fijn, R. C.; Hiemstra, D.; Phillips, R. A.; and Winden, J. V. D. (2013). "Arctic Terns Sterna paradisaea from the Netherlands migrate record distances across three oceans to Wilkes Land, East Antarctica," *Ardea*, 101(1), p. 3–12.

Chapter 6: Dead Reckoning

1. For a detailed discussion of this subject, see the author's book: *Sextant: A Voyage Guided by the Stars and the Men Who Mapped the World's Oceans* (William Collins, 2014). p. 61–90.

2. The scientific term for DR is "path integration."

3. It is in fact possible to take sights of the sun and stars through a periscope, but since this might well reveal the presence of a patrolling nuclear submarine to an enemy, some alternative navigational method was vital.

4. The scientific term is "idiothetic."

5. Twain, M. (1872). *Roughing It*. Hartford, Conn; American Publishing Company. Ch. 31.

6. Dudchenko, P. A., *Why People Get Lost* (Oxford University Press, 2010), p. 67.

7. Souman, J. L.; Frissen, I.; Sreenivasa, M. N.; and Ernst, M. O. (2009). "Walking straight into circles," *Current Biology*, 19(18), p. 1538–42.

8. Thomson, J. A. (1983). "Is continuous visual monitoring necessary in visually guided locomotion?," *Journal of Experimental Psychology: Human Perception and Performance*, 9(3), p. 427.

9. Cheung, A.; Zhang, S.; Stricker, C.; and Srinivasan, M. V. (2008). "Animal navigation: general properties of directed walks," *Biological Cybernetics*, 99(3), p. 197–217.

10. Gill, R. E.; Tibbitts, T. L.; Douglas, D. C.; Handel, C. M.; Mulcahy, D. M.; Gottschalck, J.C.; . . . and Piersma, T. (2009). "Extreme endurance flights by landbirds crossing the Pacific Ocean: ecological corridor rather than barrier?," *Proceedings of the Royal Society of London B: Biological Sciences*, 276(1656), p. 447–57.

11. Piersma, T., and Gill Jr., R. E. (1998). "Guts don't fly: small digestive organs in obese bar-tailed godwits," *The Auk*, p. 196–203.

12. Battley, P. F.; Warnock, N.; Tibbitts, T. L.; Gill, R. E.; Piersma, T.; Hassell, C. J.; . . . and Melville, D. S. (2012). "Contrasting extreme long-distance migration patterns in bar-tailed godwits *Limosa lapponica*," *Journal of Avian Biology*, 43(1), p. 21–32.

Chapter 7: The Racehorse of the Insect World

1. Wehner, R. (2013). "Life as a cataglyphologist—and beyond," *Annual Review of Entomology*, 58, p. 1–18.

2. Pfeiffer, K., and Homberg, U. (2014). "Organization and functional roles of the central complex in the insect brain," *Annual Review of Entomology*, 59, p. 165–84.

3. Wehner, R. (1987). "Matched filters—neural models of the external world," *Journal of Comparative Physiology A: Neuroethology, Sensory, Neural, and Behavioral Physiology*, 161(4), p. 511–31.

4. Srinivasan, M.; Zhang, S.; and Bidwell, N. (1997). "Visually mediated odometry in honeybees," *Journal of Experimental Biology*, 200(19), p. 2513–22.

5. Wittlinger, M.; Wehner, R.; and Wolf, H. (2006). "The ant odometer: stepping on stilts and stumps," *Science*, 312(5782), p. 1965–7.

6. Wehner, R., and Räber, F. (1979). "Visual spatial memory in desert ants, Cataglyphis bicolor (*Hymenoptera: Formicidae*)," *Experientia*, 35, p. 1569–71; Cartwright, B. A., Collett, T. S. (1983). "Landmark learning in bees:

experiments and models," *Journal of Comparative Physiology A*, 151, p. 521–43; Möller, R., and Vardy, A. (2006). "Local visual homing by matched-filter descent in image distances," *Biological Cybernetics*, 95, p. 413–30; Zeil, J., Hofmann, M. I., Chahl, J. S. (2003). "The catchment areas of panoramic snapshots in outdoor scenes," *Journal of the Optical Society of America A*, 20, p. 450–69.

7. Lambrinos, D.; Möller, R.; Labhart, T.; Pfeifer, R.; and Wehner, R. (2000). "A mobile robot employing insect strategies for navigation," *Robot and Autonomous Systems*, 30, p. 39–64.

8. Fleischmann, P. N.; Grob, R.; Müller, V. L.; Wehner, R.; and Rössler, W. (2018). "The geomagnetic field is a compass cue in cataglyphis ant navigation," *Current Biology*.

9. Shi, N. N.; Tsai, C. C.; Camino, F.; Bernard, G. D.; Yu, N.; and Wehner, R. (2015). "Keeping Cool: enhanced optical reflection and heat dissipation in silver ants," *Science*, aab3564.

10. Darwin, C., *The Descent of Man*, op. cit., pt I, p. 54.

11. Heinze, S. (2015). "Neuroethology: unweaving the senses of direction," *Current Biology*, 25(21), R1034–37.

12. See for example: Weber, K.; Venkatesh, S.; and Srinivasan, M. V. (August 1996). "Insect inspired behaviours for the autonomous control of mobile robots," *International Conference on Pattern Recognition, Proceedings, 1996*, p. 156, IEEE; Weber, K.; Venkatesh, S.; and Srinivasan, M. V. (August 1998); "An insect-based approach to robotic homing," *Fourteenth International Conference on Pattern Recognition, 1998, Proceedings*, vol. 1, p. 297–9; Expert, F.; Viollet, S.; & Ruffier, F. (2011). "Outdoor field performances of insect-based visual motion sensors," *Journal of Field Robotics*, 28(4), p. 529–41; Graham, P., and Philippides, A. (2014). "Insect-Inspired Visual Systems and Visually Guided Behavior," *Encyclopedia of Nanotechnology*, p. 1–9.

13. Collett, M., and Collett, T. S. (2018). "How does the insect central complex use mushroom body output for steering?" *Current Biology*, 28(13), R733–4.

14. Read, M. A.; Grigg, G. C.; Irwin, S. R.; Shanahan, D.; and Franklin, C. E. (2007). "Satellite tracking reveals long distance coastal travel and homing by translocated estuarine crocodiles, Crocodylus porosus," *PLOS One*, 2(9), e949.

Chapter 8: Steering by the Shape of the Sky

1. Chepesiuk, R. (2009). "Missing the dark: health effects of light pollution," *Environmental Health Perspectives*, 117 (1), A20.

2. Falchi, F.; Cinzano, P.; Duriscoe, D.; Kyba, C. C.; Elvidge, C. D.; Baugh, K.; . . . and Furgoni, R. (2016). "The new world atlas of artificial night sky brightness," *Science Advances*, 2(6), e1600377.

3. Kyba, C. C.; Kuester, T.; de Miguel, A. S.; Baugh, K.; Jechow, A.; Hölker, F.; . . . and Guanter, L. (2017). "Artificially lit surface of earth at night increasing in radiance and extent," *Science Advances*, 3(11), e1701528.

4. See for example: Stevens, R. G.; Blask, D. E.; Brainard, G. C.; Hansen, J.; Lockley, S. W.; Provencio, I.; Rea, M. S.; and Reinlib, L. (2007). "Meeting report: The role of environmental lighting and circadian disruption in cancer and other diseases," *Environmental Health Perspectives*, 115, p. 1357–62.

5. See for example: Longcore, T., and Rich, C. (2004). "Ecological light pollution," *Frontiers in Ecology and the Environment*, 2(4), p. 191–8. Also: Horváth, G.; Kriska, G.; Malik, P.; and Robertson, B. (2009). "Polarized light pollution: a new kind of ecological photopollution," *Frontiers in Ecology and the Environment*, 7(6), p. 317–25; Gaston, K. J.; Bennie, J.; Davies, T. W.; and Hopkins, J. (2013). "The ecological impacts of nighttime light pollution: a mechanistic appraisal," *Biological Reviews*, 88(4), p. 912–27.

6. For more information, visit the website of the International Dark Sky Association: darksky.org.

7. Gladwin, T., *East is a Big Bird: Navigation and Logic on Puluwat Atoll* (Harvard, 1970), p. 130–1.

8. Lewis, D., *We, the Navigators: The Ancient Art of Landfinding in the Pacific* (University of Hawaii Press, 1994, 2nd ed.), p. 94–7.

9. Gladwin, T., *East Is a Big Bird*, op. cit., p. 152.

10. Arab sailors in the Indian Ocean and Red Sea also used the "star compass" system, and it is possible that it reached them via Madagascar, which was settled by people from what is now Indonesia. See: Tolmacheva, M. (1980). "On the Arab system of nautical orientation," *Arabica*, 27 (Fasc. 2), p. 180–92.

11. Lewis, D., *We, the Navigators*, op. cit., p. 123.

12. Ibid., p. 162–3.

13. Ibid., p. 170.

14. Ibid., p. 224.

15. Gladwin, T., *East is a Big Bird*, p. 196.

16. Swan, L. W., *Tales of the Himalaya: Adventures of a Naturalist* (Mountain N' Air Books, 2000).

17. Hawkes, L. A.; Batbayar, N.; Butler, P. J.; Chua, B.; Frappell, P. B.; Meir, J. U.; . . . and Takekawa, J. Y. (2017). "Do bar-headed geese train for high altitude flights?," *Integrative and Comparative Biology*, 57(2), p. 240–51.

Chapter 9: How Birds Find True North

1. Hedenström, A.; Norevik, G.; Warfvinge, K.; Andersson, A.; Bäckman, J.; and Åkesson, S. (2016). "Annual 10-month aerial life phase in the common swift Apus apus," *Current Biology*, 26(22), p. 3066–70.

2. Aristotle, *History of Animals*, IX 49B, p. 632.

3. Clarke, W. E., *Studies in Bird Migration* (London and Edinburgh, 1912), vol. 1, p. 9–11.

4. wired.com/2014/10/fantastically-wron g-scientist-thought- birds-migrate-moon/.

5. White, G., *The Natural History of Selborne* (Folio Society, 1962), p. 102.

6. My thanks go to my nephew, Philip Morgan, for bringing the story of the arrow-stork to my attention.

7. Audubon, J. J., *The Birds of America* (New York, 1856), vol. 1, p. 227–8.

8. Kays, R.; Crofoot, M. C.; Jetz, W.; and Wikelski, M. (2015). "Terrestrial animal tracking as an eye on life and planet," *Science*, 348(6240), aaa2478.

9. Symes, C. T., and Woodborne, S. (2010). "Migratory connectivity and conservation of the Amur Falcon Falco amurensis: a stable isotope perspective," *Bird Conservation International*, 20(2), p. 134–48.

10. Anderson, R. C. (2009). "Do dragonflies migrate across the western Indian Ocean?" *Journal of Tropical Ecology*, 25(4), p. 347–58.

11. Sometimes young birds can even be persuaded to follow a human guide, flying alongside them in a microlight aircraft. This technique has been employed in efforts to preserve the critically endangered whooping crane in North America, and, more recently, to restore the northern bald ibis to its traditional breeding grounds in Europe. The spectacle of a flock of birds devotedly following a human pilot is certainly touching, though such close contact with people may well weaken the birds' ability to raise their own young successfully.

12. Willemoes, M.; Strandberg, R.; Klaassen, R. H.; Tøttrup, A. P.; Vardanis, Y.; Howey, P. W.; . . . and Alerstam, T. (2014). "Narrow-front loop migration in a population of the common cuckoo Cuculus canorus, as revealed by satellite telemetry," *PLOS One*, 9 (1), e83515.

NOTES | 271

13. Sauer, E. F., and Sauer, E. M. (January1960). "Star Navigation of Nocturnal Migrating Birds: The 1958 Planetarium Experiments," *Cold Spring Harbor Symposia on Quantitative Biology* (Cold Spring Harbor Laboratory Press), vol. 25, p. 463–73.

14. Emlen, S. T. (1967). "Migratory orientation in the indigo bunting, passerina cyanea. Pt I: Evidence for use of celestial cues," *The Auk*, 84(3), p. 309–42. See also: Emlen, S. T. (1967). "Migratory orientation in the Indigo Bunting, Passerina cyanea. Pt II: Mechanism of celestial orientation," *The Auk*, 84(4), p. 463–89.

15. Emlen, S. T. (1975). "The stellar-orientation system of a migratory bird," *Scientific American*, 233(2), p. 102–11.

16. Mouritsen, H., and Larsen, O. N. (2001). "Migrating songbirds tested in computer-controlled Emlen funnels use stellar cues for a time-independent compass," *Journal of Experimental Biology*, 204(22), p. 3855–65.

17. Strycker, N. K., *The Thing with Feathers: The Surprising Lives of Birds and What They Reveal about Being Human* (Riverhead Books, 2014).

Chapter 10: Heavenly Dung Beetles

1. Baird, E.; Byrne, M. J.; Smolka, J.; Warrant, E. J.; and Dacke, M. (2012). "The dung beetle dance: an orientation behavior?" *PLOS One*, 7(1), e30211.

2. Dacke, M.; Nilsson, D. E.; Scholtz, C. H.; Byrne, M.; and Warrant, E. J. (2003). "Animal behavior: insect orientation to polarized moonlight," *Nature*, 424 (6944), p. 33.

3. Dacke, M.; Baird, E.; Byrne, M.; Scholtz, C. H.; and Warrant, E. J. (2013). "Dung beetles use the Milky Way for orientation," *Current Biology*, 23(4), p. 298–300.

4. Sotthibandhu, S., and Baker, R. R. (1979). "Celestial orientation by the large yellow underwing moth, *Noctua pronuba L.*," *Animal Behavior*, 27, p. 786–800.

5. Ugolini, A.; Hoelters, L. S.; Ciofini, A.; Pasquali, V.; and Wilcockson, D. C. (2016). "Evidence for discrete solar and lunar orientation mechanisms in the beach amphipod, Talitrus saltator Montagu (*Crustacea, Amphipoda*)," *Scientific Reports*, 6.

6. Mauck, B.; Gläser, N.; Schlosser, W.; and Dehnhardt, G. (2008). "Harbor seals (*Phoca vitulina*) can steer by the stars," *Animal Cognition*, 11(4), p. 715–18.

7.	For a recent review of the navigational use of stars by animals, see: Foster, J. J.; Smolka, J.; Nilsson, D. E.; and Dacke, M. (January 2018). "How animals follow the stars," *Proc. R. Soc. B.*, vol. 285, no. 1871, p. 20172322, The Royal Society.

Chapter 11: Giant Peacocks

1.	*Saturnia pyri* has a wingspan of up to eight inches.

2.	Fabre, J-H., *Souvenirs Entomologique*, série VII, ch. 23. Author's translation.

3.	Farkas, S. R., and Shorey, H. H. (1972). "Chemical trail-following by flying insects: a mechanism for orientation to a distant odor source," *Science*, 178(4056), p. 67–8.

4.	Kennedy, J. S.; Ludlow, A.R.; and Sanders, C. J. (1980). "Guidance system used in moth sex attraction," *Nature*, 288(5790), p. 475–7.

5.	Martin, H. (1965). "Osmotropotaxis in the honeybee," *Nature*, 208(5005), p. 59–63.

6.	Borst, A., and Heisenberg, M. (1982). "Osmotropotaxis in Drosophila melanogaster," *Journal of Comparative Physiology A: Neuroethology, Sensory, Neural, and Behavioral Physiology*, 147(4), p. 479–84.

7.	Steck, K.; Knaden, M.; and Hansson, B. S. (2010). "Do desert ants smell the scenery in stereo?" *Animal Behavior*, 79(4), p. 939–45.

8.	Hasler, A. D., and Scholz, A. T. (2012). "Olfactory imprinting and homing in salmon: Investigations into the mechanism of the imprinting process," *Springer Science & Business Media*, vol. 14, p. xii.

9.	Nevitt, G., and Dittman, A. (1998). "A new model for olfactory imprinting in salmon," *Integrative Biology: Issues, News, and Reviews*, published in association with The Society for Integrative and Comparative Biology, 1(6), p. 215–23.

10.	Dittman, A., and Quinn, T. (1996). "Homing in Pacific salmon: mechanisms and ecological basis," *Journal of Experimental Biology*, 199(1), p. 83–91.

11.	Gatty, H., *Finding Your Way Without Map or Compass* (Dover Books, 1983), p. 32–3.

12.	Smell and taste are closely related, but rely on separate sensory organs—located in the nose and mouth respectively. In combination they yield what we perceive as "flavors." I shall concentrate here only on smell.

13.	Aristotle, *On the Soul*, II.9.

14.	Aristotle, *Sense and Sensibilia*, II.5.

15.	McGann, J. P. (2017). "Poor human olfaction is a 19th-century myth," *Science*, 356(6338), eaam7263.

16. Darwin, C., *The Descent of Man*, pt 1, p. 17–18.

17. Freud, S., *Drei Abhandlungen zur Sexualtheorie* (F. Deuticke, 1905), p. 83. Cited in McGann (2017), op. cit.

18. Bushdid, C.; Magnasco, M. O.; Vosshall, L. B.; and Keller, A. (2014). "Humans can discriminate more than 1 trillion olfactory stimuli," *Science*, 343 (6177), p. 1370–2.

19. McGann, J. P. (2017), op. cit.

20. Gottfried, J. A. (2009). "Function follows form: ecological constraints on odor codes and olfactory percepts," *Current Opinion in Neurobiology*, 19(4), p. 422–9.

21. Shepherd, G. M., *Neurogastronomy* (Columbia University Press, 2011), p. 89–90.

22. Gottfried, J. A. (2009). "Function follows form: ecological constraints on odor codes and olfactory percepts," *Current Opinion in Neurobiology*, 19(4), p. 422–9.

23. Proust, M. (trans. Scott Moncrieff, C.K. & Gilmartin, T.), *Remembrance of Things Past: Swann's Way* (Penguin, 1983), p. 48–50.

24. Shepherd, G. M., *Neurogastronomy*, op. cit., p. 111.

25. Pause, B. M. (2012). "Processing of body odor signals by the human brain," *Chemosens Percept*, 5, p. 55–63. doi: 10.1007/s12078-011-9108-2; pmid: 22448299

26. McGann, J. P. (2017). op. cit.

27. Porter, J.; Craven, B.; Khan, R. M.; Chang, S. J.; Kang, I.; Judkewitz, B.; . . . and Sobel, N. (2007). "Mechanisms of scent-tracking in humans," *Nature Neuroscience*, 10(1), p. 27–9.

28. Jacobs, L. F.; Arter, J.; Cook, A.; and Sulloway, F J. (2015). "Olfactory orientation and navigation in humans," *PLOS One*, 10(6), e0129387.

29. Rogers, L. L. (1987). "Navigation by adult black bears," *Journal of Mammalogy*, 68(1), p. 185–8.

Chapter 12: Can Birds Smell Their Way Home?

1. *Nature*, 7, (February 20, 1873), p. 303.

2. Papi, F.; Fiore, L.; Fiaschi, V.; and Benvenuti, S. (1971). "The influence of olfactory nerve section on the homing capacity of carrier pigeons," *Monitore Zoologico Italiano*, 5, p. 265–7.

3. Papi, F.; Fiore, L.; Fiaschi, V.; and Benvenuti, S. (1972). "Olfaction and homing in pigeons," *Monitore Zoologico Italiano*, 6, p. 85–95.

4. The cutting of the olfactory nerve (under general anesthetic) that connects the bird's smell receptors to its olfactory bulb, or the use of local anesthetics or caustic chemicals (like zinc sulfate) to desensitize them temporarily. The birds apparently recover very quickly from operations to sever the olfactory nerve, though they do not regain their sense of smell.

5. See for example: Benvenuti, S.; Fiaschi, V.; Fiore, L.; and Papi, F. (1973). "Homing performances of inexperienced and directionally trained pigeons subjected to olfactory nerve section," *Journal of Comparative Physiology*, 83, p. 81–92; and Biro, D.; Meade, J.; and Guilford, T. (2004). "Familiar route loyalty implies visual pilotage in the homing pigeon," *Proc. Natl. Acad. Sci. USA*, 101, p. 17440–3.

6. Baldaccini, N. E.; Benvenuti, S.; Fiaschi, V.; and Papi, F. (1975). "Pigeon navigation: effects of wind deflection at home cage on homing behaviour," *J. Comp. Physiol*, 99, p. 177–86.

7. See for example: Gagliardo, A.; Ioalè, P.; Odetti, F.; and Bingman, V.P. (2001). "The ontogeny of the homing pigeon navigational map: evidence for a sensitive learning period," *Proc. Biol. Sci*, 268, p. 197–202.

8. See for example: Phillips, J. B., and Waldvogel, J. A. (1988). "Celestial polarized light patterns as a calibration reference for sun compass of homing pigeons," *Journal of Theoretical Biology*, 131(1), p. 55–67.

9. For a detailed review, see: Gagliardo, A. (2013). "Forty years of olfactory navigation in birds," *Journal of Experimental Biology*, 216(12), p. 2165–71.

10. Wallraff, H. G. (2015). "An amazing discovery: bird navigation based on olfaction," *Journal of Experimental Biology*, 218(10), p. 1464–6.

11. Benvenuti, S., and Wallraff, H. G. (1985). "Pigeon navigation: site simulation by means of atmospheric odors," *J. Comp. Physiol. A.*, 156, p. 737–46.

12. Jorge, P. E.; Marques, A. E.; and Phillips, J. B. (2009). "Activational rather than navigational effects of odors on homing of young pigeons," *Current Biology*, 19(8), p. 650–4.

13. Gagliardo, A.; Pollonara, E.; and Wikelski, M. (2018). "Only natural local odours allow homeward orientation in homing pigeons released at unfamiliar sites," *J. Comp. Physiol. A.*, p. 1–11.

14. Walcott, C.; Wiltschko, W.; Wiltschko, R.; and Zupanc, G. K. (2018). "Olfactory navigation versus olfactory activation: a controversy revisited."

15. Nevitt, G. A. (2008). "Sensory ecology on the high seas: the odor world of the procellariiform seabirds," *Journal of Experimental Biology*, 211 (11), p. 1706–13. Similarly, the olfactory bulb of the homing pigeon—though smaller than that of a shearwater—is larger than that of a non-homing

pigeon; see: Mehlhorn, J., and Rehkämper, G. (2009). "Neurobiology of the homing pigeon—a review," *Naturwissenschaften*, 96(9), p. 1011–25.

16. Gagliardo, A.; Bried, J.; Lambardi, P.; Luschi, P.; Wikelski, M.; and Bonadonna, F. (2013). "Oceanic navigation in Cory's shearwaters: evidence for a crucial role of olfactory cues for homing after displacement," *Journal of Experimental Biology*, 216(15), p. 2798–805.

17. Pollonara, E.; Luschi, P.; Guilford, T.; Wikelski, M.; Bonadonna, F.; and Gagliardo, A. (2015). "Olfaction and topography, but not magnetic cues, control navigation in a pelagic seabird: displacements with shearwaters in the Mediterranean Sea," *Scientific Reports*, 5, srep16486.

18. Padget, O.; Bond, S. L.; Kavelaars, M. M.; van Loon, E.; Bolton, M.; Fayet, A. L.; . . . and Guilford, T. (2018). "In Situ Clock Shift Reveals that the Sun Compass Contributes to Orientation in a Pelagic Seabird," *Current Biology*.

19. Padget, O.; Dell'Ariccia, G.; Gagliardo, A.; González-Solís, J.; and Guilford, T. (2017). "Anosmia impairs homing orientation but not foraging behavior in free-ranging shearwaters," *Scientific Reports*, 7.

20. Abolaffio, M.; Reynolds, A. M.; Cecere, J. G.; Paiva, V. H.; and Focardi, S. (2018). "Olfactory-cued navigation in shearwaters: linking movement patterns to mechanisms," *Scientific Reports*, 8(1), p. 11590.

21. Debose, J. L., and Nevitt, G. A. (2008). "The use of odors at different spatial scales: comparing birds with fish," *Journal of Chemical Ecology*, 34(7), p. 867–81. doi.org/10.1007/s10886-008-9493-4.

22. Nevitt, G. A., and Bonadonna, F. (2005). "Sensitivity to dimethyl sulphide suggests a mechanism for olfactory navigation by seabirds," *Biology Letters*, 1(3), p. 303–5.

23. Mouritsen, H. (2018). "Long-distance navigation and magnetoreception in migratory animals," *Nature*, 558(7708), p. 50.

24. Benhamou, S.; Bried, J.; Bonadonna, F.; and Jouventin, P. (2003). "Homing in pelagic birds: a pilot experiment with white-chinned petrels released in the open sea," *Behavioural Processes*, 61(1–2), p. 95–100; Bonadonna, F.; Bajzak, C.; Benhamou, S.; Igloi, K.; Jouventin, P.; Lipp, H.P.; and Dell'Omo, G. (2005). "Orientation in the wandering albatross: interfering with magnetic perception does not affect orientation performance," *Proceedings of the Royal Society of London B: Biological Sciences*, 272(1562), p. 489–95.

25. Mora, C. V.; Davison, M.; Wild, J. M.; and Walker, M. M. (2004). "Magnetoreception and its trigeminal mediation in the homing pigeon," *Nature*, 432(7016), p. 508.

26. Wallraff, H. G. (1980). "Does pigeon homing depend on stimuli perceived during displacement?," *Journal of Comparative Physiology A: Neuroethology, Sensory, Neural, and Behavioral Physiology*, 139(3), p. 193–201.

27. See for example: Wiltschko, R., and Wiltschko, W. (2017). "Considerations on the role of olfactory input in avian navigation," *Journal of Experimental Biology*, 220(23), p. 4347–50.

28. Guilford, T.; Freeman, R.; Boyle, D.; Dean, B.; Kirk, H.; Phillips, R.; and Perrins, C. (2011). "A dispersive migration in the Atlantic puffin and its implications for migratory navigation," *PLOS One*, 6(7), e21336.

Chapter 13: Sound Navigation

1. Gatty, H., *Finding Your Way Without Map or Compass* (Dover Books, 1983), p. 78–9.

2. Konishi, M. (1993). "Listening with two ears," *Scientific American*, 268(4), p. 66–73.

3. Wilson, C., "Human bat uses echoes and sounds to see the world," *New Scientist*, May 6, 2015.

4. Flanagin, V.L.; Schörnich, S.; Schranner, M.; Hummel, N.; Wallmeier, L.; Wahlberg, M.; . . . and Wiegrebe, L. (2017). "Human exploration of enclosed spaces through echolocation," *Journal of Neuroscience*, 37(6), p. 1614–27. See also: Thaler, L.; Reich, G. M.; Zhang, X.; Wang, D.; Smith, G. E.; Tao, Z.; et al. (2017). "Mouth-clicks used by blind expert human echolocators—signal description and model-based signal synthesis," *PLOS Comput Biol.*, 13(8), e1005670.

5. Balcombe, J., *What a Fish Knows: The Inner Lives of Our Underwater Cousins* (Scientific American/Farrar, Straus and Giroux, 2016), p. 44.

6. Kemp, C., "The original batman," *New Scientist*, November 15, 2017.

7. Griffin, D. R.; Webster, F. A.; and Michael, C. R. (1960). "The echolocation of flying insects by bats," *Animal Behavior*, 8(3–4), p. 141–54.

8. Barn owls too can find their prey in the dark using just their ears. They can detect the faint sounds made by mice or voles as they scuttle through the grass and pinpoint their location with amazing precision.

9. Ulanovsky, N., and Moss, C. F. (2008). "What the bat's voice tells the bat's brain," *Proceedings of the National Academy of Sciences*, 105(25), p. 8491–8.

10. Waterman, T. H., *Animal Navigation* (Scientific American Library, 1989), p. 131–3.

11. Verfuß, U. K.; Miller, L. A.; and Schnitzler, H. U. (2005). "Spatial orientation in echolocating harbor porpoises (*Phocoena phocoena*)," *Journal of Experimental Biology*, 208(17), p. 3385–94.

12. Kreithen, M. L., and Quine, D. B. (1979). "Infrasound detection by the homing pigeon: a behavioral audiogram," *Journal of Comparative Physiology*, 129(1), p. 1–4.

13. I often heard the very loud, double "boom-boom" of Concorde when out at sea in the middle of the English Channel.

14. Hagstrum, J. T. (2000). "Infrasound and the avian navigational map," *Journal of Experimental Biology*, 203(7), p. 1103–11.

15. Grant, U. S. (1895), *Personal Memoirs of U. S. Grant.*, Sampson Low, ch. 28. For further examples, see: nellaware.com/blog/acoustic-shadow-in-the-civil-war.html.

16. Hagstrum, J. T. (2013). "Atmospheric propagation modeling indicates homing pigeons use loft-specific infrasonic 'map' cues," *Journal of Experimental Biology*, 216(4), p. 687–99.

17. Quine, D. B., and Kreithen, M. L. (1981). "Frequency shift discrimination: Can homing pigeons locate infrasounds by Doppler shifts?," *Journal of Comparative Physiology*, 141(2), p. 153–5.

18. Wallraff, H. G. (1972). "Homing of pigeons after extirpation of their cochleae and lagenae," *Nature*, 236(68), p. 223–4.

19. Hagstrum, J. T., and Manley, G. A. (2015). "Releases of surgically deafened homing pigeons indicate that aural cues play a significant role in their navigational system," *Journal of Comparative Physiology A*, 201(10), p. 983–1001.

20. Hagstrum, J. T.; McIsaac, H. P.; and Drob, D. P. (2016). "Seasonal changes in atmospheric noise levels and the annual variation in pigeon homing performance," *Journal of Comparative Physiology A*, 202(6), p. 413–24.

21. Hoffman, J. I., and Forcada, J. (2012). "Extreme natal philopatry in female Antarctic fur seals (*Arctocephalus gazelle*)," *Mammalian Biology-Zeitschrift für Säugetierkunde*, 77(1), p. 71–3.

Chapter 14: The Earth's Magnetism

1. For a full discussion, see: Taylor, E. G. R., *The Haven-Finding Art: A History of Navigation from Odysseus to Captain Cook* (Hollis and Carter, 1956), ch. 5.

2. In fact, it is the interaction between the liquid outer core and the mysterious primordial magnetic field of the inner core that generates the geomagnetic field. My thanks to Jon Hagstrum for pointing this out to me.

3. The lines of force emerge from the north pole of a magnet. Confusingly, in the case of Earth, this happens to be the magnetic pole located closest to the geographic south pole.

4. Sailors use instead the term "magnetic variation," perhaps to avoid confusion with *celestial* declination—one of the key parameters used in celestial navigation.

5. For a good visualization, see: maps.ngdc.noaa.gov/viewers/historical_declination.

6. For maps showing how magnetic declination, inclination and intensity vary across the earth's surface, see the website of the US National Oceanic and Atmospheric Administration: ngdc.noaa.gov/geomag/WMM/image.shtml.

7. For such a map, see: ngdc.noaa.gov/geomag/WMM/data/WMM2015/WMM2015_F_MERC.pdf.

8. Viguier, C. (1882). "Le sens de l'orientation et ses organes chez les animaux et chez l'homme," *Revue Philosophique de la France et de l'Etranger*, p. 1–36.

9. Gould, J. L., and Gould, C. G., *Nature's Compass* (Princeton University Press, 2012), p. 100–4.

10. Merkel, F. W., and Wiltschko, W. (1965), "Magnetismus und richtungsfinden zugunruhiger rotkehlchen (Erithacus rubeculaj)," *Vogelwarte*, 23(1), p. 71–77.

11. Wiltschko, W., and Wiltschko, R. (1972). "Magnetic compass of European robins," *Science*, 176(4030), p. 62–4.

12. Able, K. P., and Able, M. A. (1993). "Daytime calibration of magnetic orientation in a migratory bird requires a view of skylight polarization," *Nature*, 364(6437), p. 523.

13. Cochran, W. W.; Mouritsen, H.; and Wikelski, M. (2004). "Migrating song-birds recalibrate their magnetic compass daily from twilight cues," *Science*, 304(5669), p. 405–8.

14. Wiltschko, W., and Wiltschko, R. (2005). "Magnetic orientation and magnetoreception in birds and other animals," *Journal of Comparative Physiology A*, 191(8), p. 675–93.

15. Bottesch, M.; Gerlach, G.; Halbach, M.; Bally, A.; Kingsford, M.J.; and Mouritsen, H. (2016). "A magnetic compass that might help coral reef fish larvae return to their natal reef," *Current Biology*, 26(24), R1266–7.

16. Phillips, J. B., and Sayeed, O. (1993). "Wavelength-dependent effects of light on magnetic compass orientation in Drosophila melanogaster," *Journal of Comparative Physiology A: Neuroethology, Sensory, Neural, and Behavioral Physiology*, 172(3), p. 303–8.

17. Vácha, M.; Drštková, D.; and Pužová, T. (2008). "Tenebrio beetles use magnetic inclination compass," *Naturwissenschaften*, 95(8), p. 761–5.

18. Rasmussen, K.; Palacios, D. M.; Calambokidis, J.; Saborío, M. T.; Dalla Rosa, L.; Secchi, E. R.; . . . and Stone, G. S. (2007). "Southern Hemisphere humpback whales wintering off Central America: insights from water temperature into the longest mammalian migration," *Biology Letters*, 3(3), p. 302–5.

19. Horton, T. W.; Holdaway, R. N.; Zerbini, A. N.; Hauser, N.; Garrigue, C.; Andriolo, A.; and Clapham, P. J. (2011). "Straight as an arrow: humpback whales swim constant course tracks during long-distance migration," *Biology Letters*, rsbl20110279.

20. Bailey, H.; Senior, B.; Simmons, D.; Rusin, J.; Picken, G.; and Thompson, P. M. (2010). "Assessing underwater noise levels during pile-driving at an offshore windfarm and its potential effects on marine mammals," *Marine Pollution Bulletin*, 60(6), p. 888–97.

21. Kirschvink, J. L.; Dizon, A. E.; and Westphal, J. A. (1986). "Evidence from strandings for geomagnetic sensitivity in cetaceans," *Journal of Experimental Biology*, 120(1), p. 1–24; and Kirschvink, J. L., "Geomagnetic sensitivity in cetaceans: an update with live stranding records in the United States," *Sensory Abilities of Cetaceans* (Boston, MA: Springer, 1990), p. 639–49.

22. Vanselow, K. H.; Jacobsen, S.; Hall, C.; and Garthe, S. (2017). "Solar storms may trigger sperm whale strandings: explanation approaches for multiple strandings in the North Sea in 2016," *International Journal of Astrobiology*, p. 1–9.

23. Garrigue, C.; Clapham, P. J.; Geyer, Y.; Kennedy, A. S.; Zerbini, A. N. (2015). "Satellite tracking reveals novel migratory patterns and the importance of seamounts for endangered South Pacific humpback whales," *Royal Society Open Science*, 2, 150489: dx.doi.org/10.1098/rsos.150489.

Chapter 15: So How Does the Monarch Navigate?

1. For a review of the early history of the monarch migration puzzle, see: Brower, L. (1996). "Monarch butterfly orientation: missing pieces of a magnificent puzzle," *Journal of Experimental Biology*, 199(1), p. 93–103.

2. Urquhart, F., *The Monarch Butterfly* (University of Toronto Press, 1960), p. viii.

3. Ibid.

4. The following account of the monarch migration relies heavily on: Walbauer, G. (2000). *Millions of Monarchs, Bunches of Beetles: How Bugs Find Strength in Numbers* (Harvard University Press), p. 50–70.

5. Barker, J. F., and Herman, W. S. (1976). "Effect of photoperiod and temperature on reproduction of the monarch butterfly, Danaus plexippus," Journal of Insect Physiology, 22(12), p. 1565–8.

6. Perez, S. M.; Taylor, O. R.; and Jander, R. (1997). "A sun compass in monarch butterflies," *Nature*, 387(6628), p. 29.

7. Mouritsen, H., and Frost, B. J. (2002). "Virtual migration in tethered flying monarch butterflies reveals their orientation mechanisms," *Proceedings of the National Academy of Sciences*, 99(15), p. 10162–6.

8. This technique is described in more detail in chapter 17.

9. Reppert, S. M.; Zhu, H.; and White, R. H. (2004). "Polarized light helps monarchs migrate," *Current Biology*, 14(2), p. 155–8.

10. Merlin, C.; Gegear, R. J.; and Reppert, S. M. (2009). "Antennal circadian clocks coordinate sun compass orientation in migratory monarch butterflies," *Science*, 325 (5948), p. 1700–4. See also: Guerra, P. A.; Merlin, C.; Gegear, R. J.; and Reppert, S. M. (2012). "Discordant timing between antennae disrupts sun compass orientation in migratory monarch butterflies," *Nature Communications*, 3, p. 958.

11. Heinze, S., and Reppert, S. M. (2011). "Sun compass integration of skylight cues in migratory monarch butterflies," *Neuron*, 69(2), p. 345–58.

12. Guerra, P. A.; Gegear, R. J.; and Reppert, S. M. (2014). "A magnetic compass aids monarch butterfly migration," *Nature Communications*, 5.

13. Reppert, S. M.; Guerra, P. A.; and Merlin, C. (2016). "Neurobiology of monarch butterfly migration," *Annual Review of Entomology*, 61, p. 25–42.

14. Stalleicken, J.; Mukhida, M.; Labhart, T.; Wehner, R.; Frost, B. J.; and Mouritsen, H. (2005). "Do monarch butterflies use polarized skylight for orientation?" *Journal of Experimental Biology*, 208, p. 2399–408.

15. Mouritsen, H.; Derbyshire, R.; Stalleicken, J.; Mouritsen, O. Ø.; Frost, B. J.; and Norris, D. R. (2013). "An experimental displacement and over 50 years of tag-recoveries show that monarch butterflies are not true navigators," *Proceedings of the National Academy of Sciences*, 110(18), p. 7348–53.

16. Anderson, R. C. (2009). "Do dragonflies migrate across the western Indian Ocean?" *Journal of Tropical Ecology*, 25(4), p. 347–58.

17. Hobson, K. A.; Anderson, R. C.; Soto, D. X.; and Wassenaar, L. I. (2012). "Isotopic evidence that dragonflies (*Pantala flavescens*) migrating through the Maldives come from the northern Indian subcontinent," *PLOS One*, 7(12), e52594.

18. Chapman, J. W.; Reynolds, D. R.; and Wilson, K. (2015). "Long-range seasonal migration in insects: mechanisms, evolutionary drivers and ecological consequences," *Ecology Letters*, 18(3), p. 287–302.

Chapter 16: The Silver "Y"

1. Nesbit, R. L.; Hill, J. K.; Woiwod, I. P.; Sivell, D.; Bensusan, K. J.; and Chapman, J. W. (2009). "Seasonally adaptive migratory headings mediated by a sun compass in the painted lady butterfly, Vanessa cardui," *Animal Behavior*, 78(5), p. 1119–25.

2. Chapman, J. W.; Bell, J. R.; Burgin, L. E.; Reynolds, D. R.; Pettersson, L. B.; Hill, J. K.; . . . and Thomas, J. A. (2012). "Seasonal migration to high latitudes results in major reproductive benefits in an insect," *Proceedings of the National Academy of Sciences*, 109(37), p. 14924–9.

3. Hu, G.; Lim, K. S.; Horvitz, N.; Clark, S. J.; Reynolds, D. R.; Sapir, N.; and Chapman, J. W. (2016). "Mass seasonal bioflows of high-flying insect migrants," *Science*, 354(6319), p. 1584–7.

4. Chapman, J. W., et al. (2010). "Flight orientation behaviors promote optimal migration trajectories in high-flying insects," *Science*, 327, p. 682–5.

5. Gaston, A. J.; Hashimoto, Y.; and Wilson, L. (2015). "First evidence of east–west migration across the North Pacific in a marine bird," *Ibis*, 157(4), p. 877–82.

Chapter 17: The Dark Lord of the Snowy Mountains

1. There are other populations of bogong in Australia that migrate in different directions.

2. Warrant, E.; Frost, B.; Green, K.; Mouritsen, H.; Dreyer, D.; Adden, A.; . . . and Heinze, S. (2016). "The Australian Bogong moth Agrotis infusa: a long-distance nocturnal navigator," *Frontiers in Behavioral Neuroscience*, 10.

3. Heinze, S.; and Warrant, E. (2016). "Bogong moths," *Current Biology*, 26(7), R263–5

4. Ibid.

5. Quoted in: Warrant, E.; Frost, B.; Green, K.; Mouritsen, H.; Dreyer, D.; Adden, A.; . . . and Heinze, S. (2016). "The Australian Bogong moth Agrotis infusa: a long-distance nocturnal navigator," *Frontiers in Behavioral Neuroscience*, 10.

6. Dreyer, D.; Frost, B.; Mouritsen, H.; Günther, A.; Green, K.; Whitehouse, M.; . . . and Warrant, E. (2018). "The Earth's Magnetic Field and Visual Landmarks Steer Migratory Flight Behavior in the Nocturnal Australian Bogong Moth," *Current Biology*.

7. Pittman, S. E.; Hart, K. M.; Cherkiss, M. S.; Snow, R. W.; Fujisaki, I.; Smith, B. J.; . . . and Dorcas, M. E. (2014). "Homing of invasive Burmese pythons in South Florida: evidence for map and compass senses in snakes," *Biology Letters*, 10(3), 20140040.

Chapter 18: Map and Compass Navigation

1. The technical terms are "allocentric" and "egocentric" respectively.

2. It is sometimes also called "true navigation."

3. Two signals would not be enough as their circles would cross in two different places—thereby creating ambiguity.

4. Perdeck, A. C. (1958). "Two Types of Orientation in Migrating Starlings, *Sturnus vulgaris L.*, and Chaffinches, *Fringilla coelebs L.*, as Revealed by Displacement Experiments," *Ardea*, 46(1–2), p. 1–2.

5. Schmidt-Koenig, K., and Schlichte, H. J. (1972). "Homing in pigeons with impaired vision," *Proceedings of the National Academy of Sciences*, 69(9), p. 2446–7; and Schmidt-Koenig, K., and Walcott, C. (1978). "Tracks of pigeons homing with frosted lenses," *Animal Behavior*, 8(26), p. 480–6.

6. Walcott, C., and Schmidt-Koenig, K. (1973). "The effect on pigeon homing of anesthesia during displacement," *The Auk* 3, 90, p. 281–6.

7. Wallraff, H. G. (2013). "Ratios among atmospheric trace gases together with winds imply exploitable information for bird navigation: a model elucidating experimental results," *Biogeosciences*, 10(11), p. 6929–43.

8. Wallraff, H. (2005). "Beyond familiar landmarks and integrated routes: goal-oriented navigation by birds," *Connection Science*, 17(1–2), p. 91–106.

9. Boström, J. E.; Åkesson, S.; and Alerstam, T. (2012). "Where on earth can animals use a geomagnetic bi-coordinate map for navigation?" *Ecography*, 35(11), p. 1039–47.

10. For fuller discussion, see: Mouritsen, H. (2013). "The Magnetic Senses," in: Galizia, C. G.; Lledo, P.-M. (eds.), *Neurosciences—From Molecule to Behavior: A University Textbook*, DOI 10.1007/978-3-642-10769-6_20, p. 427–43.

11. Muheim, R. (2011). "Behavioural and physiological mechanisms of polarized light sensitivity in birds," *Philosophical Transactions of the Royal Society of London B: Biological Sciences*, 366(1565), p. 763–71.

12. Waterman, T. H. (2006). "Reviving a neglected celestial underwater polarization compass for aquatic animals," *Biological Reviews*, 81(1), p. 111–15.

13. Powell, S. B.; Garnett, R.; Marshall, J.; Rizk, C.; and Gruev, V. (2018). "Bioinspired polarization vision enables underwater geolocalization," *Science Advances*, 4(4), eaao6841.

Chapter 19: Can Birds Solve the Longitude Problem?

1. Thorup, K.; Bisson, I.-A.; Bowlin, M. S.; Holland, R. A.; Wingfield, J. C.; Ramenofsky, M.; and Wikelski, M. (2007). "Evidence for a navigational map stretching across the continental U.S. in a migratory songbird," *Proc. Natl. Acad. Sci. USA*, 104, p. 18115–19.

2. Chernetsov, N.; Kishkinev, D.; and Mouritsen, H. (2008). "A long-distance avian migrant compensates for longitudinal displacement during spring migration," *Current Biology*, 18(3), p. 188–90.

3. Piggins, H. D., and Loudon, A. (2005). "Circadian biology: clocks within clocks," *Current Biology*, 15(12), R455–7.

4. Kishkinev, D.; Chernetsov, N.; and Mouritsen, H. (2010). "A Double-Clock or Jetlag Mechanism is Unlikely to be Involved in Detection of East–West Displacements in a Long-Distance Avian Migrant," *The Auk*, 127(4), p. 773–80.

5. Kishkinev, D.; Chernetsov, N.; Pakhomov, A.; Heyers, D.; and Mouritsen, G. (2015). "Eurasian reed warblers compensate for virtual magnetic displacement," *Current Biology*, 25(19), R822–4.

6. Kishkinev, D.; Chernetsov, N.; Heyers, D.; and Mouritsen, H. (2013). "Migratory reed warblers need intact trigeminal nerves to correct for a 1,000 km eastward displacement," *PLOS One*, 8(6), e65847.

7. Chernetsov, N.; Pakhomov, A.; Kobylkov, D.; Kishkinev, D.; Holland, R.A.; and Mouritsen, H. (2017). "Migratory Eurasian reed warblers can use magnetic declination to solve the longitude problem," *Current Biology*, 27(17), p. 2647–51.

8. Quinn, T. P., and Brannon, E. L. (1982). "The use of celestial and magnetic cues by orienting sockeye salmon smolts," *J. Comp. Physiol.*, 147, p. 547–52.

9. Putman, N. F.; Lohmann, K. J.; Putman, E. M.; Quinn, T. P.; Klimley, A. P.; and Noakes, D. L. (2013). "Evidence for geomagnetic imprinting as a homing mechanism in Pacific salmon," *Current Biology*, 23(4), p. 312–16.

10. Putman, N. F.; Scanlan, M. M.; Billman, E. J.; O'Neil, J. P.; Couture, R. B.; Quinn, T. P.; . . . and Noakes, D. L. (2014). "An inherited magnetic map guides ocean navigation in juvenile Pacific salmon," *Current Biology*, 24(4), p. 446–50.

11. Obleser, P.; Hart, V.; Malkemper, E. P.; Begall, S.; Holá, M.; Painter, M. S.; . . . and Burda, H. (2016). "Compass-controlled escape behavior in roe deer," *Behavioral Ecology and Sociobiology*, 70(8), p. 1345–55.

Chapter 20: The Mystery of Sea Turtle Navigation

1. Carr, A. F., *The Sea Turtle* (University of Texas, 1986), p. 26–7.

2. Ibid., p. 159.

3. Ibid., p. 163–5.

4. Papi, F.; Liew, H. C.; Luschi, P.; and Chan, E. H. (1995). "Long-range migratory travel of a green turtle tracked by satellite: evidence for navigational ability in the open sea," *Marine Biology*, 12(2), p. 171–5.

5. Luschi, P.; Papi, F.; Liew, H. C.; Chan, E. H.; and Bonadonna, F. (1996). "Long-distance migration and homing after displacement in the green turtle (Chelonia mydas): a satellite tracking study," *Journal of Comparative Physiology A*, 178(4), p. 447–52.

6. Papi, F.; Luschi, P.; Crosio, E.; and Hughes, G. R. (1997). "Satellite tracking experiments on the navigational ability and migratory behaviour of the loggerhead turtle Caretta caretta," *Marine Biology*, 129(2), p. 215–20.

7. Hughes, G. R.; Luschi, P.; Mencacci, R.; and Papi, F. (1998). "The 7000-km oceanic journey of a leatherback turtle tracked by satellite," *Journal of Experimental Marine Biology and Ecology*, 229(2), p. 209–17.

8. Luschi, P.; Åkesson, S.; Broderick, A. C.; Glen, F.; Godley, B. J.; Papi, F.; and Hays, G. C. (2001). "Testing the navigational abilities of ocean migrants: displacement experiments on green sea turtles (*Chelonia mydas*)," *Behavioral Ecology and Sociobiology*, 50(6), p. 528–34.

9. Hays, G. C.; Åkesson, S.; Broderick, A. C.; Glen, F.; Godley, B. J.; Papi, F.; and Luschi, P. (2003). "Island-finding ability of marine turtles," *Proceedings of the Royal Society of London B: Biological Sciences*, 270 (suppl. 1), S5–7.

10. Luschi, P.; Benhamou, S.; Girard, C.; Ciccione, S.; Roos, D.; Sudre, J.; and Benvenuti, S. (2007). "Marine turtles use geomagnetic cues during open-sea homing," *Current Biology*, 17(2), p. 126–33.

11. Bonanomi, S.; Overgaard Therkildsen, N.; Retzel, A.; Berg Hedeholm, R.; Pedersen, M.W.; Meldrup, D.; . . . and Nielsen, E. E. (2016). "Historical DNA documents long-distance natal homing in marine fish," *Molecular Ecology*, 25(12), p. 2727–34.

Chapter 21: Costa Rican Adventures

1. The website for Lohmann's lab provides a good overview of their research, including helpful graphics and access to many of their publications: unc.edu/depts/oceanweb/turtles/.

2. Lohmann, K. J., and Lohmann, C. M. (1992), "Orientation to waves by green turtle hatchlings." *Journal of Experimental Biology*, 171(1) p. 1–13.

3. Link to video clip: unc.edu/depts/oceanweb/turtles/.

4. Stewart, B. S., and DeLong, R. L. (1995). "Double migrations of the northern elephant seal, *Mirounga angustirostris*," *Journal of Mammalogy*, 76(1), p. 196–205.

5. Bonfil, R.; Meÿer, M.; Scholl, M. C.; Johnson, R.; O'Brien, S.; Oosthuizen, H.; . . . and Paterson, M. (2005). "Transoceanic migration, spatial dynamics, and population linkages of white sharks," *Science*, 310(5745), p. 100–3.

6. Anderson, J. M.; Clegg, T. M.; Véras, L. V.; and Holland, K. N. (2017). "Insight into shark magnetic field perception from empirical observations," *Scientific Reports*, 7(1), p. 11042.

7. Horton, T. W.; Hauser, N.; Zerbini, A. N.; Francis, M. P.; Domeier, M. L.; Andriolo, A.; . . . and Holdaway, R. N. (2017). "Route Fidelity During Marine Megafauna Migration," *Frontiers in Marine Science*, 4, p. 422.

Chapter 22: A Light in the Darkness

1. Link to description of magnetic coil system: unc.edu/depts/oceanweb/turtles.

2. Lohmann, K. J., and Lohmann, C. M. (1994). "Detection of magnetic inclination angle by sea turtles: a possible mechanism for determining latitude," *Journal of Experimental Biology*, 194(1), p. 23–32.

3. Lohmann, K. J.; Lohmann, C. M. F.; Ehrhart, L. M.; Bagley, D. A., and Swing, T. (2004). "Geomagnetic map used in sea-turtle navigation," *Nature*, 428, p. 909–10.

4. Putman, N. F., and Mansfield, K. L. (2015). "Direct evidence of swimming demonstrates active dispersal in the sea turtle 'lost years,'" *Current Biology*, 25(9), p. 1221–7.

5. Lohmann, K. J., and Lohmann, C. M. (1996). "Detection of magnetic field intensity by sea turtles," *Nature*, 380(6569), p. 59.

6. The hatchlings became disoriented when "sent" to a location well outside the gyre: Fuxjager, M. J.; Eastwood, B. S.; and Lohmann, K. J. (2011). "Orientation of hatchling loggerhead sea turtles to regional magnetic fields along a transoceanic migratory pathway," *Journal of Experimental Biology*, 214(15), p. 2504–8.

7. Lohmann, K. J.; Cain, S. D.; Dodge, S. A.; and Lohmann, C. M. (2001). "Regional magnetic fields as navigational markers for sea turtles," *Science*, 294(5541), p. 364–6.

8. Putman, N. F.; Verley, P.; Endres, C. S.; and Lohmann, K. J. (2015). "Magnetic navigation behavior and the oceanic ecology of young loggerhead sea turtles," *Journal of Experimental Biology*, 218(7), p. 1044–50.

9. Summarized in: Lohmann, K. J.; Putman, N. F.; and Lohmann, C. M. (2012). "The magnetic map of hatchling loggerhead sea turtles," *Current Opinion in Neurobiology*, 22(2), p. 336–42.

10. Putman, N. F.; Endres, C. S.; Lohmann, C. M.; and Lohmann, K. J. (2011). "Longitude perception and bicoordinate magnetic maps in sea turtles," *Current Biology*, 21(6), p. 463–6.

11. Putman, N. F., and Lohmann, K. J. (2008). "Compatibility of magnetic imprinting and secular variation," *Current Biology*, 18(14), R596–7.

12. Brothers, J. R., and Lohmann, K. J. (2015). "Evidence for geomagnetic imprinting and magnetic navigation in the natal homing of sea turtles," *Current Biology*, 25(3), p. 392–6.

13. Brothers, J. R., and Lohmann, K. J. (2018). "Evidence that Magnetic Navigation and Geomagnetic Imprinting Shape Spatial Genetic Variation in Sea Turtles," *Current Biology*, 28(8), p. 1325–9.

14. Endres, C. S., and Lohmann, K. J. (2013). "Detection of coastal mud odors by loggerhead sea turtles: a possible mechanism for sensing nearby land," *Marine Biology*, 160(11), p. 2951–6.

15. Endres, C. S.; Putman, N. F.; Ernst, D. A.; Kurth, J. A.; Lohmann, C. M.; and Lohmann, K. J. (2016). "Multi-modal homing in sea turtles: modeling dual use of geomagnetic and chemical cues in island-finding," *Frontiers in Behavioral Neuroscience*, 10, p. 19.

16. Lohmann, K. J.; Lohmann, C. M.; and Endres, C. S. (2008). "The sensory ecology of ocean navigation," *Journal of Experimental Biology*, 211(11), p. 1719–28.

17. Lohmann, K.; Pentcheff, N.; Nevitt, G.; Stetten, G.; Zimmer-Faust, R.; Jarrard, H.; and Boles, L. C. (1995). "Magnetic orientation of spiny lobsters in the ocean: experiments with undersea coil systems," *Journal of Experimental Biology*, 198(10), p. 2041–8.

18. Boles, L. C., and Lohmann, K. J. (2003). "True navigation and magnetic maps in spiny lobsters," *Nature*, 421(6918), p. 60–3.

19. Baker, R. R. (1980). "Goal orientation by blindfolded humans after long-distance displacement: Possible involvement of a magnetic sense," *Science*, 210(4469), p. 555–7.

20. Fildes, B. N.; O'Loughlin, B. J.; Bradshaw, J. L.; and Ewens, W. J. (1984). "Human orientation with restricted sensory information: no evidence for magnetic sensitivity," *Perception*, 13(3), p. 229–48.

21. In July 2018.

22. Naisbett-Jones, L. C.; Putman, N. F.; Stephenson, J.F.; Ladak, S.; and Young, K. A. (2017). "A magnetic map leads juvenile European eels to the Gulf Stream," *Current Biology*, 27(8), p. 1236–40.

23. Durif, C. M.; Bonhommeau, S.; Briand, C.; Browman, H. I.; Castonguay, M.; Daverat, F.; . . . and Moore, A. (2017). "Whether European eel leptocephali use the earth's magnetic field to guide their migration remains an open question," *Current Biology*, 27(18), R998–1000.

Chapter 23: The Great Magnetic Mystery

1. Kobayashi, A., and Kirschvink, J. L. (1995). "Magnetoreception and electromagnetic field effects: sensory perception of the geomagnetic field in animals and humans."

2. Taylor, B. K.; Johnsen, S.; and Lohmann, K. J. (2017). "Detection of magnetic field properties using distributed sensing: a computational neuroscience approach," *Bioinspiration & Biomimetics*, 12(3), 036013.

3. Gould, J. L., and Gould, C. G., *Nature's Compass*, op. cit., p. 111–14.

4. Anderson, J. M.; Clegg, T. M.; Véras, L. V.; and Holland, K. N. (2017). "Insight into shark magnetic field perception from empirical observations," *Scientific Reports*, 7(1), p. 11042.

5. Fleissner, G.; Stahl, B.; Thalau, P.; Falkenberg, G.; and Fleissner, G. (2007). "A novel concept of Fe-mineral-based magnetoreception: histological and physicochemical data from the upper beak of homing pigeons," *Naturwissenschaften*, 94(8), p. 631–42.

6. Mora, C. V.; Davison, M.; Wild, J. M.; and Walker, M. M. (2004). "Magnetoreception and its trigeminal mediation in the homing pigeon," *Nature*, 432(7016), p. 508.

7. Treiber, C. D.; Salzer, M. C.; Riegler, J.; Edelman, N.; Sugar, C.; Breuss, M.; . . . and Shaw, J. (2012). "Clusters of iron-rich cells in the upper beak of pigeons are macrophages not magnetosensitive neurons," *Nature*, 484(7394), p. 367.

8. Zapka, M.; Heyers, D.; Hein, C. M.; Engels, S.; Schneider, N. L.; Hans, J.; . . . and Mouritsen, H. (2009). "Visual but not trigeminal mediation of magnetic compass information in a migratory bird," *Nature*, 461(7268), p. 1274.

9. Gagliardo, A.; Ioalè, P.; Savini, M.; and Wild, J. M. (2006). "Having the nerve to home: trigeminal magnetoreceptor versus olfactory mediation of homing in pigeons," *Journal of Experimental Biology*, 209(15), p. 2888–92.

10. Kishkinev, D.; Chernetsov, N.; Heyers, D.; and Mouritsen, H. (2013). "Migratory reed warblers need intact trigeminal nerves to correct for a 1,000 km eastward displacement," *PLOS One*, 8(6), e65847.

11. Holland, R. A., and Helm, B. (2013). "A strong magnetic pulse affects the precision of departure direction of naturally migrating adult but not juvenile birds," *Journal of The Royal Society Interface*, 10(81), 20121047.

12. For a detailed review, see: Mouritsen, H. (2015). "Magnetoreception in birds and its use for long-distance migration," *Sturkie's Avian Physiology*, p. 113–33.

13. Wu, L. Q., and Dickman, J. D. (2012). "Neural correlates of a magnetic sense," *Science*, 336(6084), p. 1054–7.

14. Schulten, K.; Swenberg, C. E.; and Weller, A. (1978). "A biomagnetic sensory mechanism based on magnetic field modulated coherent electron spin motion," *Zeitschrift für Physikalische Chemie*, 111(1), p. 1–5.

15. For a detailed review of the evidence relating to radical pairs, see: Hore, P. J., and Mouritsen, H. (2016). "The radical-pair mechanism of magnetoreception," *Annual Review of Biophysics*, 45, p. 299–344.

16. Zapka, M.; Heyers, D.; Hein, C. M.; Engels, S.; Schneider, N. L.; Hans, J.; . . . and Mouritsen, H. (2009). "Visual but not trigeminal mediation of magnetic compass information in a migratory bird," *Nature*, 461(7268), p. 1274.

17. Gegear, R. J.; Casselman, A.; Waddell, S.; and Reppert, S. M. (2008). "Cryptochrome mediates light-dependent magnetosensitivity in Drosophila," *Nature*, 454(7207), p. 1014. See also: Gegear, R. J.; Foley, L. E.; Casselman, A.; and Reppert, S. M. (2010). "Animal cryptochromes mediate magnetoreception by an unconventional photochemical mechanism," *Nature*, 463(7282), p. 804.

18. Bazalova, O.; Kvicalova, M.; Valkova, T.; Slaby, P.; Bartos, P.; Netusil, R.; . . . and Damulewicz, M. (2016). "Cryptochrome 2 mediates directional magnetoreception in cockroaches," *Proceedings of the National Academy of Sciences*, 113(6), p. 1660–5.

19. Jungerman, R. L., and Rosenblum, B. (1980). "Magnetic induction for the sensing of magnetic fields by animals—an analysis," *Journal of Theoretical Biology*, 87(1), p. 25–32.

20. Lauwers, M.; Pichler, P.; Edelman, N. B.; Resch, G. P.; Ushakova, L.; Salzer, M. C.; . . . and Keays, D. A. (2013). "An iron-rich organelle in the cuticular plate of avian hair cells," *Current Biology*, 23(10), p. 924–9.

21. Nordmann, G. C.; Hochstoeger, T.; and Keays, D. A. (2017). "Magnetoreception—a sense without a receptor," *PLOS Biology*, 15(10), e2003234.

22. Tawa, A.; Ishihara, T.; Uematsu, Y.; Ono, T.; and Ohshimo, S. (2017). "Evidence of westward transoceanic migration of Pacific bluefin tuna in the Sea of Japan based on stable isotope analysis," *Marine Biology*, 164(4), p. 94. See also: Block, B. A., et al. (2005). "Electronic tagging and population structure of Atlantic bluefin tuna," *Nature* 434, p. 1121–7.

23. Willis, J.; Phillips, J.; Muheim, R.; Diego-Rasilla, F. J.; and Hobday, A. J. (2009). "Spike dives of juvenile southern bluefin tuna (*Thunnus maccoyii*): a navigational role?" *Behavioral Ecology and Sociobiology*, 64(1), p. 57.

24. Walker, M. M. (1984). "Learned magnetic field discrimination in yellowfin tuna, *Thunnus albacares*," *Journal of Comparative Physiology A: Neuroethology, Sensory, Neural, and Behavioral Physiology*, 155(5), p. 673–9.

Chapter 24: The Seahorses in Our Heads

1. De Waal, F., *Are We Smart Enough to Know How Smart Animals Are?* (Granta, 2016), p. 55.

2. Tolman, E. C. (1948). "Cognitive maps in rats and men," *Psychological Review*, 55(4), p. 189.

3. Summarized in: Gould, J. L., and Gould, C. G., *Nature's Compass*, p. 155–7.

4. Gazzaniga, M. S.; Ivry, R. B.; and Mangun, G. R., *Cognitive Neuroscience* (W. W. Norton, 2002), p. 18.

5. See for example: Hubel, D. H., and Wiesel, T. N. (1963). "Shape and arrangement of columns in cat's striate cortex," *The Journal of Physiology*, 165(3), p. 559–68.

6. Similar "lobectomies" are still widely undertaken, but with much more care and precision to remove diseased tissue that is thought to be the origin of the epilepsy.

7. Scoville, W. B., and Milner, B. (1957). "Loss of recent memory after bilateral hippocampal lesions," *Journal of Neurology, Neurosurgery, and Psychiatry*, 20(1), p. 11.

8. O'Keefe, J., and Dostrovsky, J. (1971). "The hippocampus as a spatial map. Preliminary evidence from unit activity in the freely moving rat," *Brain Research*, 34(1), p. 171–5.

9. O'Keefe, J., and Nadel, L., *The Hippocampus as a Cognitive Map* (Oxford University Press, 1978).

10. Fyhn, M.; Molden, S.; Witter, M. P.; Moser, E. I.; and Moser, M. B. (2004). "Spatial representation in the entorhinal cortex," *Science*, 305(5688),

p. 1258–64. See also: Hafting, T.; Fyhn, M.; Molden, S.; Moser, M. B.; and Moser, E. I. (2005). "Microstructure of a spatial map in the entorhinal cortex," *Nature*, 436(7052), p. 801.

11. For an up-to-date catalog, see: Grieves, R. M., and Jeffery, K. J. (2017). "The representation of space in the brain," *Behavioral Processes*, 135, p. 113–31.

12. It is a rule that a Nobel Prize cannot be awarded to more than three people.

13. Sherry, D. F.; Grella, S. L.; Guigueno, M. F.; White, D. J.; and Marrone, D. F. (2017). "Are There Place Cells in the Avian Hippocampus?," *Brain, Behavior and Evolution*, 90(1), p. 73–80.

14. Geva-Sagiv, M.; Las, L.; Yovel, Y.; and Ulanovsky, N. (2015). "Spatial cognition in bats and rats: from sensory acquisition to multiscale maps and navigation," *Nature Reviews Neuroscience*, 16(2), p. 94.

15. Finkelstein, A.; Las, L.; Ulanovsky, N. (2016). "3-D maps and compasses in the brain," *Annual Review of Neuroscience*, 39, p. 171–96. See also: Grieves, R. M., and Jeffery, K. J. (2017). "The representation of space in the brain," *Behavioral Processes*, 135, p. 113–31.

16. Ulanovsky, N., and Moss, C. F. (2007). "Hippocampal cellular and network activity in freely moving echolocating bats," *Nature Neuroscience*, 10(2), p. 224–33.

17. Eichenbaum, H., and Cohen, N. J. (2014). "Can we reconcile the declarative memory and spatial navigation views on hippocampal function?," *Neuron*, 83(4), p. 764–70.

18. Moser, E. I.; Moser, M. B.; and McNaughton, B. L. (2017). "Spatial representation in the hippocampal formation: a history," *Nature Neuroscience*, 20(11), pp, 1448–64.

19. Buzsáki, G., and Llinás, R. (2017). "Space and time in the brain," *Science*, 358(6362), p. 482–5.

20. Pašukonis, A.; Loretto, M. C.; and Hödl, W. (2017). "Map-like navigation from distances exceeding routine movements in the three-striped poison frog (*Ameerega trivittata*)," *Journal of Experimental Biology*, jeb-169714.

Chapter 25: The Human Navigational Brain

1. Hort, J.; Laczó, J.; Vyhnálek, M.; Bojar, M.; Bureš, J.; and Vlček, K. (2007). "Spatial navigation deficit in amnestic mild cognitive impairment," *Proceedings of the National Academy of Sciences*, 104(10), p. 4042–7.

2. See for example: niallmclaughlin.com/projects/alzheimers-respite-centre-dublin.

3. Maguire, E. A.; Gadian, D. G.; Johnsrude, I. S.; Good, C. D.; Ashburner, J.; Frackowiak, R. S.; and Frith, C. D. (2000). "Navigation-related structural change in the hippocampi of taxi drivers," *Proceedings of the National Academy of Sciences*, 97(8), p. 4398–403.

4. Interestingly though, there seemed to be a trade-off. The front part of the hippocampus in the controls was larger than that in the taxi drivers, which may mean that the taxi drivers are less adept at recalling some kinds of visual information.

5. Maguire, E. A.; Woollett, K.; and Spiers, H. J. (2006). "London taxi drivers and bus drivers: a structural MRI and neuropsychological analysis," *Hippocampus*, 16(12), p. 1091–101.

6. Konishi, K., and Bohbot, V. D. (2013). "Spatial navigational strategies correlate with gray matter in the hippocampus of healthy older adults tested in a virtual maze," *Frontiers in Aging Neuroscience*, 5.

7. Stern, Y. (2006). "Cognitive reserve and Alzheimer disease," *Alzheimer Disease & Associated Disorders*, 20, S69–74. See also: Xu, W.; Yu, J. T.; Tan, M. S.; and Tan, L. (2015). "Cognitive reserve and Alzheimer's disease," *Molecular Neurobiology*, 51(1), p. 187–208.

8. Epstein, R. A.; Patai, E. Z.; Julian, J. B.; and Spiers, H. J. (2017). "The cognitive map in humans: spatial navigation and beyond," *Nature Neuroscience*, 20(11), p. 1504.

9. Rubin, R. D.; Watson, P. D.; Duff, M. C.; and Cohen, N. J. (2014). "The role of the hippocampus in flexible cognition and social behavior," *Frontiers in Human Neuroscience*, 8, p. 742.

10. Kuehn, E.; Chen, X.; Geise, P.; Oltmer, J.; and Wolbers, T. (2018). "Social targets improve body-based and environment-based strategies during spatial navigation," *Experimental Brain Research*, p. 1–10.

11. Omer, D. B.; Maimon, S. R.; Las, L.; and Ulanovsky, N. (2018). "Social place-cells in the bat hippocampus," *Science*, 359(6372), p. 218–24. See also: Danjo, T.; Toyoizumi, T.; and Fujisawa, S. (2018). "Spatial representations of self and other in the hippocampus," *Science*, 359(6372), p. 213–18. See also: Okuyama, T.; Kitamura, T.; Roy, D. S.; Itohara, S.; and Tonegawa, S. (2016). "Ventral CA1 neurons store social memory," *Science*, 353(6307), p. 1536–41.

12. Beadle, J. N.; Tranel, D.; Cohen, N.J.; and Duff, M. (2013). "Empathy in hippocampal amnesia," *Frontiers in Psychology*, 4, p. 69.

13. Tavares, R. M.; Mendelsohn, A.; Grossman, Y.; Williams, C. H.; Shapiro, M.; Trope, Y.; and Schiller, D. (2015). "A map for social navigation in the human brain," *Neuron*, 87(1), p. 231–43.

14. Vashro, L., and Cashdan, E. (2015). "Spatial cognition, mobility, and reproductive success in northwestern Namibia," *Evolution and Human Behavior*, 36(2), p. 123–9.

15. Duff, M. C.; Kurczek, J.; Rubin, R.; Cohen, N. J.; and Tranel, D. (2013). "Hippocampal amnesia disrupts creative thinking," *Hippocampus*, 23(12), p. 1143–9.

16. Warren, D. E.; Kurczek, J.; and Duff, M. C. (2016). "What relates newspaper, definite, and clothing? An article describing deficits in convergent problem solving and creativity following hippocampal damage," *Hippocampus*, 26(7), p. 835–40.

17. Constantinescu, A. O.; O'Reilly, J. X.; and Behrens, T. E. (2016). "Organizing conceptual knowledge in humans with a gridlike code," *Science*, 352(6292), p. 1464–68.

18. Coutrot, A.; Silva, R.; Manley, E.; de Cothi, W.; Sami, S.; Bohbot, V.; . . . and Spiers, H. (2017). *Global determinants of navigation ability*. Current Biology, 28(17), 2861–2866. The app can be downloaded from the internet: seaheroquest.com/site/en.

19. Polansky, L.; Kilian, W.; and Wittemyer, G. (April 2015). "Elucidating the significance of spatial memory on movement decisions by African savannah elephants using state–space models," *Proc. R. Soc. B.*, vol. 282, no. 1805, p. 20143042, The Royal Society.

20. Schmitt, M. H.; Shuttleworth, A.; Ward, D.; and Shrader, A. M. (2018). "African elephants use plant odours to make foraging decisions across multiple spatial scales," *Animal Behavior*, 141, p. 17–27.

Chapter 26: The Language of the Earth

1. Levi, P. (trans. Wolf, S.), *The Truce* (Abacus, 1987), p. 349–51.

2. Solnit, R., *A Field Guide to Getting Lost* (Canongate, 2006), p. 10.

3. Carr, N. (2013). "All can be lost: The risk of putting our knowledge in the hands of machines," *The Atlantic*, 11, p. 1–12.

4. Parasuraman, R., and Manzey, D. H. (2010). "Complacency and bias in human use of automation: An attentional integration," *Human Factors*, 52(3), p. 381–410.

5. telegraph.co.uk/news/earth/countryside/9090729/Warning-over-decline-i n-map-skills-as-ramblers-rely-on-sat-navs.html.

6. Iaria, G., and Barton, J. J. (2010). "Developmental topographical disorientation: a newly discovered cognitive disorder," *Experimental Brain Research*, 206(2), p. 189–96.

7. Aporta, C., et al. (2005). *Current Anthropology*, 46(5), p. 729–53.

8. Carr, N. (2013). *The Atlantic*, 11, p. 1–12.

9. Hemingway, E., *The Sun Also Rises* (Scribner, 1926), ch. 13, p. 136.

10. pressherald.com/2016/05/25/report-geraldine-largay-kept-journal-durin g-weeks-lost-in-maine-woods/document.

Chapter 27: Conclusions

1. Balbuena, M. S.; Tison, L.; Hahn, M.-L.; Greggers, U.; Menzel, R.; and Farina, W. M. (2015). "Effects of sublethal doses of glyphosate on honeybee navigation," *The Journal of Experimental Biology*, 218, p. 2799–805. doi:10.1242/jeb.117291.

2. For more information, see: International Dark Sky Association (darksky. org).

3. Cited in: Singer, P., *Animal Liberation* (Random House, 1990), p. 192.

4. Aquinas, T., *Summa Contra Gentiles*, bk 3, pt 2, ch. 112.

5. Aristotle, *Politics*, bk 1, ch. 8.

6. See for example: newyorker.com/news/daily-comment/ are-evangelica l-leaders-saving-scott-pruitts-job.

7. Wilson, E. O., *Biophilia: The Human Bond with Other Species* (Harvard, 1984), p. 85.

8. aeon.co/essays/why-forests-and-rivers-are-the-most-potent-health-tonic-around.

9. Kuo, M. (2015). "How might contact with nature promote human health? Promising mechanisms and a possible central pathway," *Frontiers in Psychology*, 6, p. 1093.

10. Piff, P. K.; Dietze, P.; Feinberg, M.; Stancato, D. M.; and Keltner, D. (2015). "Awe, the small self, and prosocial behavior," *Journal of Personality and Social Psychology*, 108(6), p. 883.

Index

M

magnetic navigation
bacteria, 6–7, 213
bees, 125–26, 213–14
birds, 104, 126–27, 170–77,
214–15, 216–18
eels, 211
fish, 177–79, 214, 217–19
geomagnetic field, 120–25
humans, 209–10
insects, 134–35, 141–42,
144–54, 213–14
lobsters, 206–09
newts, 8
sensors for, 212–19
turtles, 196–97, 199–206
worms, 8
Magnus, Olaus, 75
Maguire, Eleanor, 232
Mann, Richard, 26
Manx shearwater(s), 80–81
map and compass navigation
animal use of, 161–64
birds, 165–68, 170–77
crustaceans, 169
Mazzeo, Rosario, 81
Merkel, Friedrich, 126
Middendorf, Alexander von,
125
migration of birds, 75–80
Miller, George, 223
Mills, Enos, 92–93, 243
Molaison, Henry, 224–25, 231
monarch butterflies, 4–5,
129–36, 250
Mortensen, Hans Christian, 77
Morton, Charles, 75–76

Moser, Edvard, 226–27
Moser, May-Britt, 226–27
moth(s)
bogong moth(s), 143–54
giant peacock moth(s), 89–90
silver "Y" moth(s), 3, 138–
142
mountains, 74
Mouritsen, Henrik, 133, 135,
143, 146, 149, 168, 176, 215,
217
murrelet(s), 142

N

Nadel, Lynn, 225
newt(s), 8
Nilsson, Dan-Eric, 9

O

Odyssey (Homer), 70
O'Keefe, John, 225, 227
orientation, 28

P

Papi, Floriano, 87, 98, 99–101,
112, 186–88
Pardi, Leo, 87
Pašukonis, Andreas, 229–30
pee-wee flycatcher(s), 76–77
Perdeck, Christiaan, 164–66,
176
Perez, Sandra, 132–33
Pero, 9–10
pigeon(s)
landmarks, 26
magnetic navigation, 214
map and compass navigation,
166–68, 176–77